Interop

ALSO BY JOHN PALFREY AND URS GASSER

Born Digital
Intellectual Property Strategy

Interop

The Promise and Perils of
Highly Interconnected Systems

John Palfrey

and

Urs Gasser

BASIC BOOKS
A MEMBER OF THE PERSEUS BOOKS GROUP
New York

Books published by Basic Books are available at special discounts for bulk purchases in the
United States by corporations, institutions, and other organizations. For more information,
please contact the Special Markets Department at the Perseus Books Group, 2300 Chestnut
Street, Suite 200, Philadelphia, PA 19103, or call (800) 810-4145, ext. 5000, or e-mail
special.markets@perseusbooks.com.

Library of Congress Cataloging-in-Publication Data
Palfrey, John G. (John Gorham), 1972–
  Interop : the promise and perils of highly interconnected systems / John Palfrey and Urs
Gasser.
    p. cm.
  Includes bibliographical references and index.
  ISBN 978-0-465-02197-0 (hardcover) — ISBN 978-0-465-02907-5 (e-book) 1. Systems
integration. 2. Standards, Engineering. 3. Technology—Social aspects. 4. Internetworking
(Telecommunication) I. Gasser, Urs. II. Title. III. Title: Promise and perils of highly inter-
connected systems.
  TA168.P26 2012
  620.001'1—dc23
                              2012002593
10 9 8 7 6 5 4 3 2 1

To the Berkman Center team,
whose members teach us about human interoperability every day.

CONTENTS

Why Interop Matters

T he Internet has made the world at once a smaller and a more complex place. Digital technologies connect billions of people, businesses, organizations, and governments with each other in ways that enhance our lives but that we don't fully understand. We are interconnected as never before, to our enormous benefit: we stay in touch with far-away family and friends for low cost; we learn about news instantaneously, access knowledge remotely, collaborate more efficiently, and do all kinds of business online. Our most complex systems—government agencies, financial institutions, transportation infrastructures, health care and energy systems—are linked by these new, invisible information channels, which are essential components of today's global economy.

This capacity for connection is about more than just making our lives more convenient or efficient. Organizations can become more specialized, better at what they do, and more able to collaborate effectively across distance and time—whether in business, public life, or civic activism—in ways that are changing the course of history. Consider what a highly connected network of people in the Middle East, many of them very young, were able to accomplish, with little in the way of central coordination, during the Arab spring of 2011. They toppled long-standing regimes, one after another, through peaceful activism that was powered by their high degree of digital connectedness.

But this growing level of interconnectedness comes at an increasingly high price. We make big trade-offs as we become digitally connected everywhere and anytime. We struggle to keep up with overflowing e-mail inboxes; we feel overwhelmed by the flood of news and information coming at us from all directions. We are also more vulnerable, in ways that are less obvious and less well understood. The same infrastructure that enables us to create, store, and share information can put our privacy and security at risk. Data breaches and privacy invasions make the news daily, illustrating what can happen when massive amounts of data are exchanged among complex systems without adequate safeguards. These risks are omnipresent in the digital age. They touch every aspect of modern life through the information exchanged with banks, credit card companies, mobile phone carriers, tax authorities, entertainment giants, or online businesses.

The problems of too much interconnectivity present enormous challenges both for organizations and for society at large. Our most advanced systems and infrastructures have become so complex that they are hard to manage effectively. Our financial system, for instance, has fallen into deep crisis due in part to the new vulnerabilities, complexities, and domino effects resulting from unprecedented digital connectivity. Our technological networks are so pervasive, and we use them so intensely, that we have good reason to worry that data about ourselves and our families might float out of our sight and our control. In such cases, the problem is not the interconnectivity itself but, rather, the fact that it is not adequately checked or managed.

In other cases, we suffer from too little connectivity. We struggle, for instance, to reform our health care system in no small part because we cannot get our information systems to work together properly with one another. Higher degrees of connectivity and information sharing among our health care providers would make the health care system vastly more efficient and effective at providing care—and yet that connectivity eludes us. If the devices we use and the households we live in could "talk" to each other, we could dramatically reduce our energy consumption by creating a "smart grid" that would allow for efficient energy supply. More often than not, our future success in addressing the big societal challenges of our time, from health care to climate change, will depend heavily on our ability to create better interfaces and connections among complex systems and our ability to share information appropriately.

This challenge—creating better, more useful connectivity while simultaneously finding better ways to manage its inherent risks—inspired this book. As societies, we have rushed to build information and communications infrastructures that enhance connectivity and enable the flow of information among individuals, organizations, and systems. But we have not yet developed a normative theory identifying what we want out of all this interconnectivity. We call this theory *interoperability,* or *interop.* The payoff from our theory of interop is that it can help us decide where we need interconnectedness in complex systems and at what level—and where we don't. Without such a theory, we lack a stable framework for figuring out how to harness the benefits of the digital technologies that connect us while still protecting our core societal values. And we have not yet refined a sense of which tools will get us to optimal levels of interop. This book has been written to help meet these urgent challenges—challenges that are at once highly conceptual and deeply practical.

The main purpose of the theory of interoperability is to help define the optimal level of interconnectedness and to lay out a path for achieving it. As a first step, we must develop a new lens for analyzing how complex systems, components, and applications are connected—or sometimes, inexplicably, still separated. Second, we must take a deep look at the promises and the drawbacks that come with increased connectivity. We need to balance the costs and the benefits of the connectivity we create, both in the short and long terms. We can then assess how much interconnectedness we should aim to achieve among our institutions, systems, and peoples. Most important, a theory of interoperability leads to a clear understanding of the mechanisms—technical, organizational, or legal—through which interop can be achieved and shows how we might optimize the interoperability levels of complex systems.

This book explores in depth two interop problems. The first is to figure out how to define and get to an optimal level of interoperability in complex systems. The second is to deal with the adverse effects of interoperability: loss of diversity, increasingly pressing concerns about its effects on individual privacy and security, and the risk of locking in older technologies and hindering innovation. An enormous amount hangs in the balance. Our economies, our personal well-being, and our environment will all be affected by whether these two interop problems can be solved in our most critical complex systems.

We, the authors, have been studying interoperability through a variety of methods for the past decade. We are both law professors and researchers, interested in the way the ongoing explosion of information technologies affects societies around the world. Our joint research project started out as a transatlantic collaboration. One team was based in Europe, at a leading research center at the University of St. Gallen in Switzerland. The other was based at the Berkman Center for Internet and Society at Harvard University. Our methodology is based on the development and analysis of a broad range of case studies. We started out with a series of cases that relate primarily to information and communications technologies, which is our core field of study. We have talked to hundreds of people and hosted workshops on three continents with experts in a wide range of fields.

As we got deeper and deeper into the topic, we began to see the reach of interop beyond the context of our core field. We began to research fields outside our own: economics, business, systems theory, psychology. Through our case studies, drafted by members of an interdisciplinary team of researchers at our respective centers, we began to examine areas farther afield where interop matters greatly.

Some of the biggest challenges of the age are in fact interop problems. Consider health care reform, which relies upon getting interoperability right in the context of electronic health records, or climate change, which turns in part on the emergence of a next-generation energy delivery system, the smart grid. We have posted both of these case studies online, free for anyone to read, as a companion to this book.[1] We present them as the raw data from which we have built the theory of interop and the practical suggestions that we offer in this book. Our idea, in the spirit of transparency, is that anyone can look at the data from which we have drawn our conclusions, and we hope to provoke dialogue on these pressing issues.

Much of our research has involved conducting in-depth interviews and convening workshops with experts in the fields of computing, law, and psychology, as well as in many fields of industry. Over the many years of conducting interviews on this topic, we have never found a single person who thinks that interop is anything other than a good thing in general. That is the starting point: people generally want higher levels of interconnection. After that, there is not a lot of common ground. There is no single, agreed-upon definition of interoperability. There are many views about what interop is and how it should be achieved. And there are even more views about how, if at all, the problems to which interop gives rise should be addressed.

There is no one-size-fits-all definition of interoperability. In the most general sense, in the context of information technologies, interoperability is the ability to transfer and render useful data and other information across systems, applications, or components. But it is important to go beyond this core understanding to explore a broader understanding of what interop means in different contexts and at different levels.

In theoretical terms, interoperability functions on four broad layers of complex systems. Understanding this structure is essential to understanding how interop works and how society ought to go about achieving (or thwarting) it. Interop is not just about the flow of data or about technology; it involves essential questions of human and institutional interaction as well. The problems associated with interop are just as much about culture as they are about technology.

The first layer is technological. Think of the hardware and the code in computing systems or the train tracks in the transportation systems. Interoperability at this layer means that, in the most basic sense, the systems can connect to one another, often through an explicit, agreed-upon interface.

The second layer of interoperability is the data layer. This layer is closely paired with the technology layer; indeed, the two are often inextricably linked. It is not enough for the technological systems to be able to exchange signals or to pass material from one to the next. If the receiving party cannot understand the data, then the technological interoperability is worthless. Imagine that you receive an e-mail with an attachment on your smartphone. When you click on the attachment, you get an error message: you can't read that attachment on your device. In this case, the software on your smartphone can receive the message but cannot render the data useful to you.

The third layer of interoperability is the human layer. This layer is much more abstract than the technology and data layers, but it is very important to the success of interoperability. It is one thing for the e-mail systems to exchange messages between them and for the data to be passed successfully across those interoperable systems, but it is another thing for the humans at either side of the exchange of information to understand each other and to act upon that exchange. Language is one way to think about the human layer of interoperability—in order to communicate, we need a common language—but that is only the starting point. We also need to be able to work together in other ways. Interop often succeeds or fails based on whether we are willing to put effort into working together as human beings.

At the highest and most abstract layer, we consider interoperability at the institutional layer. Just as it is essential that people work together, it is also frequently important that societal systems engage effectively. The legal system is one example of an institutional layer of interoperability (or its absence). For instance, if two companies in different countries want to collaborate, they must share a common understanding of, say, contract law. Likewise, if two companies want to start a joint venture, they need a shared understanding of corporate law. This does not mean that the two countries need to have identical legal systems or that the two companies need to adopt the same internal processes or rules. They only need to have *enough* in common that the interests of both are protected. Interoperability at the institutional layer makes possible high levels of collaboration and exchange without making the parties identical.

Given the importance of each of these four layers, no short definition of interop ends up being particularly satisfying. Interoperability is highly context specific. And so, rather than aiming for a single definition that can apply across different sectors and cases, we consider the specific contours of the interoperability at work in each example, across the four layers; we operate pragmatically and with an open working definition. This process approach to defining interoperability is meant to avoid prejudging the best way to accomplish interoperability. It is also intended to reflect the idea that interop is not a binary concept. There are degrees and types of interop, which fall along a multidimensional spectrum.

Interop also means different things to different people. The kind of interop that matters to computer users—whether an e-mail comes through the system legibly, for example—may be different from the kind of interop that matters to the Internet service providers who have to send the messages, to the companies that make the software and hardware that make the systems work in the first place, and to the police who from time to time want to be able to intercept those e-mails in order to apprehend a criminal. In the context of signing up for a new social network online, interoperability might mean being able to sign into one program or website (such as Twitter or Facebook) and having personal information seamlessly and securely transferred as needed to a variety of merchants and service providers (such as the mobile apps foursquare or SCVNGR). In the context of online music, recording industry executives might view interoperability as being able to sell their content securely through a variety of online channels and have it play on many approved devices. Web service and mashup platform providers care about seamless data transmission and easy extension and integration of data sources by users and small developers.[2]

The point is that different people and firms will have different perspectives on what interoperability means in a given context, how much interoperability is optimal, and how it ought to be accomplished. The incentives related to interoperability can vary greatly. Some firms will seek to use interoperability to keep people within their systems; others will want to profit directly or indirectly by enabling others to innovate on the basis of an open, broadly interoperable platform.

Interoperability should be an explicit goal in national and international discussions of business, law, and policy because the upsides of interoperability are massive: it fosters innovation and competition, enhances diversity, gives consumers choice, and can lead to unexpected benefits over time. Interoperability is not an end in itself; rather, it is a means to accomplish other societal goals, such as growing the economy, fighting climate change, and improving the quality of health care. Our goal should be to harness the great potential of interoperability while avoiding some of its possible downsides.

Interop can help many people in many contexts. For instance, consumers who want to be able to choose from a broad range of applications for their home entertainment systems are well advised to purchase a system that offers interoperability across different providers and services. Entrepreneurs who seek to develop and market their own web application are usually more likely to succeed if they pursue an interop-based approach. Business executives should usually strive for interoperability among teams, work flows, and the like within their organizations. Government agencies operate at lower costs and with greater efficiency, and thus can provide better service to citizens, when they (and their systems) work together. When legislators and policy makers are creating or adjusting legal frameworks aimed at fostering innovation, they should consider the various approaches described in this book to create incentives for increasing both technical and institutional interoperability.

Our approach to interoperability takes several forms. Although the most obvious context for the argument about the benefits of interop is the information and communications technology sector, interoperability matters in sectors throughout the economy. That said, although we have studied historical examples—transportation and finance in particular—to glean insights into how society has built successful interoperable systems in the past, we focus here primarily on debates that rage today in the information and communications technologies sector. These debates relate to the distribution of digital music and movies, document formats, and the long-term preservation of human knowledge. We scan the horizon for issues that are just emerging, such as cloud computing,[3] the smart grid, e-health records, and online identity systems. And we make a series of arguments about how interoperability might be achieved through law, policy, technology, and innovations in the marketplace. As we look ahead, we contend that interop-related challenges will only grow harder to manage as our systems grow more complex and interconnected.

Historical perspective is an important starting point for our study of interop. Systems have failed to work together since time immemorial. And when they have succeeded, humans have sometimes had to live with unforeseen and unwanted consequences.

The American rail system is one example of how people have worked together to solve interop problems. In May 1869, in the middle of the United States, a small group witnessed the ceremony of the golden spike, a major event in the history of interoperability. The witnesses celebrated the connection of the railroad systems, which now reached from the Pacific Ocean to the Atlantic. The golden spike, driven into the earth at the center of the country, made it possible for a train to connect the two great oceans of the world. Civilizations could be joined in a new way from one part of the globe to another.[4]

On its simplest level, the connection of train tracks from one ocean to another, across the massive North American continent, is a story about a technology. The technology of train tracks and engines and cars is an essential infrastructure in a modern economy. This technology was developed not by one single government or one firm. It was paid for and built by a whole lot of people with many different financial, political, and social interests. Those many interests were not necessarily aligned. But somehow, a system emerged that made it possible to travel at high speed from the edge of one ocean to another. The "somehow" was a commitment to make a system that could interoperate. The idea was not to develop a single train system that was the same in every respect everywhere or that was owned by the same people; rather, the goal was to achieve one that would work together across different owners and different plans for usage.

Interoperability of the rail system in the late nineteenth century in America made many other good things possible. This technology carried goods harvested or manufactured in one corner of the country to others, making new markets accessible. The railroad was also an early communications network. Over its tracks rode people, ideas, and cultural norms. The interconnected, interoperable railroad system made possible a newer, faster way for people to communicate, for markets to grow more complex and profitable, and for cultures to become connected to one another. The interoperability that the US rail system made possible reached far beyond the ability to convey trains from one place to another.

The twentieth century is full of examples that illustrate the importance of interoperability as a driver of innovation, growth, and benefits to consumers. The further development of the transportation infrastructure is one such example. Consider the ease with which people can travel by air or car across the countries of Europe, for instance, and the number of systems that need to work together to make such seamless—and safe—travel possible. Financial systems are an equally instructive example: the extent to which currency can flow from one jurisdiction to another has driven international trade and cross-cultural exchange of many sorts.

In neither of these large-scale examples—transportation and finance—has interop put an end to diversity. Systems must have sufficient overlap to work together, but they do not need to be completely standardized. This key distinction—between sameness and interoperability—recurs throughout the examples we explain in this book. In the best cases, even while systems and people are enabled to work together, the powerful force of diversity can be preserved. The point was not that there needed to be a single train company or a single bank that everyone had to use. Nor did all the train companies or banks have to do everything the same way. They just had to agree to do *some* things in ways that would interoperate. Crucially, the things they agreed to do in an interoperable manner had to be the *right* things.

Think of the trains themselves. The sizes and shapes of the trains could vary significantly from one company to another. The ways they hired and staffed their train systems could be quite different. The policies for what a given company would carry across the tracks, and how, could be widely diverse. But the gauge of the tracks and the distance between the wheels of the trains, along with a few other technical specifications, had to be the same.

Currency is another example of the compatibility between diversity and interoperability. The Swiss franc coexists with the euro and the British pound in the regional economy of Europe. Each of these currencies coexists with the US dollar, the Chinese yuan, and dozens of other widely used currencies. And yet a global economy has emerged whereby people from each of these jurisdictions can trade together without major hassles. Enabling this trade was not a process of standardizing on a single currency, with a single value and a single governor. The process has been more subtle than that, emerging from the bottom up over a long period of time and in turn enabling local diversity while giving rise to a global system of finance. The system has been made to interoperate through the establishment of intermediaries, rules, and laws.

One of the primary benefits of interoperability is that it can preserve key elements of *diversity* while ensuring that systems work together in the ways that matter most. One of the tricks to the creation of interoperable systems is to determine what the optimal level of interoperability is: in what ways should the systems work together, and in what ways should they not?

The benefits of interoperability are vast. In particular, interoperable systems make all our lives easier. Interoperable systems can make us more efficient by lowering the costs of switching between and among varying tasks. They can afford consumers more choice by limiting the effects of being locked in to any one system. They can promote cross-cultural understanding, the free movement of ideas, and the flow of trade. They can support a competitive environment for businesses. And they can often lead to innovation in the marketplace.

And yet interop is not always an unalloyed good. The breakdown in the global financial system in 2008 and the subsequent crisis stemming from Greece's defaults in 2010 illustrate the dangers of highly interoperable systems. The economies of Europe, in particular, are tightly interconnected: the European Union is an economic unit by design. The downside of this degree of interconnection is that Greece's debt woes have meant that Germany, France, and other economically stronger countries have had to foot the bill for much of Greece's overspending. The European Union, in turn, has become deeply linked to the economies of the United States and many big Asian markets. As a result, the effects of Greece's ongoing problems have been felt in every economy in the world. We have become very good at connecting our economies, but not especially good at isolating the problems that arise in one part of the world from the rest.

For a much simpler, prosaic example of a situation in which you want significant but not complete interoperability, consider a car trip. You are driving home to New York from a visit to Boston. You're tired and bleary-eyed. You also realize that you're short on gas: the little red gas indicator has lit up beside the odometer. You decide it would be a good idea to pull off the highway before driving much further.

At the next gas station you see, you put your credit card into the computer attached to the gas pump and lift the nozzle. You try to put the nozzle into your car's gas tank, but it won't go. You try again. It still won't go. You realize, to your surprise, that the nozzle is the wrong size. After a few more tries, and a curse or two, it dawns on you that you've made a mistake. Your car takes ordinary fuel, but the nozzle in your hand is connected to the tank of diesel fuel. You pick up another nozzle, corresponding to the correct kind of fuel, and soon you're on your way with a full tank.

Several things had to work together to ensure that you got your gas without ruining your engine. A complex financial system enabled you to enter your credit card information into a computing system associated with the pump at the gas station. After a flurry of bits made their way to your bank and back, the company selling the gas decided it was safe to let the gas flow into your car, because the funds would be released by the credit card issuer. And years earlier, the manufacturer of your car sized the receptacle for gas correctly so as to allow the nozzle dispensing ordinary unleaded fuel to fit properly. The gas station had to offer nozzles that fit the corresponding apertures, and so forth. Interop is the secret to these complex systems working together to enable you to get gas efficiently and safely into your car.

At the same time, the system was set up so that you couldn't put the wrong fuel—in this case, diesel, which would have harmed your engine—in your car. Likewise, if you had introduced a stolen credit card, the financial system would have denied you the fuel. The system was designed to not interoperate when it wasn't meant to. In this case, the system was meant to correct for human error (your bleary-eyed reach for the diesel nozzle) and to prevent cheating (someone's attempt to spend another person's money on fuel).

We do not always want things to interoperate completely. Sometimes we want brakes on interoperability to correct against human error, as the diesel gas pump example demonstrates (purposeful noninteroperability). Other times, we want brakes to prevent fraud, as with the example of the credit card: interoperability is blocked if there is a possibility the card is stolen. We want to make sure that the parts of the system can always work together but also that the system can throw up roadblocks or speed bumps where necessary (limited, or conditional, interoperability).

The same principle holds true with all sorts of other complex systems. In the environment of the web, we want the system to be able to pass data from one place to another, but we also want it to be able to include brakes that stop the wrong kind of personal information (for instance, health-related data) from flowing from one place to another in the wrong cases. In the global economic context, one might wish to establish firewalls that could rise up to block the effects of crisis in one market (say, Iceland or Greece or, more dangerous, China or the United States) from spreading to another (any of the other two hundred or so countries in the world). Sometimes the places where interoperability doesn't exist are as important as the places where it does.

T he debate about how to get to optimal levels of interoperability too often operates at the extremes. At one extreme, people argue that the state should have no involvement whatsoever in achieving interop. This theory is part and parcel of the dominant strain of cyberlibertarianism, which views any state involvement as anathema to innovation and to the positive development of information and technology systems. At the other end of the spectrum lies the notion that the state ought to drive the development of important complex systems by mandating certain approaches to interop. In this interventionist vision, the leadership of the state is necessary to accomplish high levels of interop; without it, the results will be too uneven and inconsistent to serve the public well.

We, as societies, should not favor one approach or another to interop in the abstract. The type of intervention we choose and who we think should lead it will vary based on a wide range of factors. For instance, when it comes to setting rules for emergency communications, the state ought to be responsible for leading the approach to interop. The state is best positioned to look out for the public's overriding interest in safety and security; also, the type of standard that needs to be set is straightforward. When it comes to determining the best way for e-mail systems to talk to one another, though, there is not much argument against the private sector's lead in terms of setting and managing the standards. In such highly technical cases, the state is ill equipped to make judgments as to standards; that expertise resides primarily in the private sector. Most cases call for a mix of approaches: interop problems tend to be more complex than either of these two simple examples.

To understand the possible options, throughout this book we map a range of approaches that fall along two broad spectrums: private-sector-led approaches ("non-regulatory approaches") versus government-driven measures ("regulatory approaches") on the one hand, and unilateral versus collaborative approaches on the other.

The chart lists the most important interop tools that we have identified in the course of our research into a broad range of examples in the information and communications industry. One way to accomplish interoper-

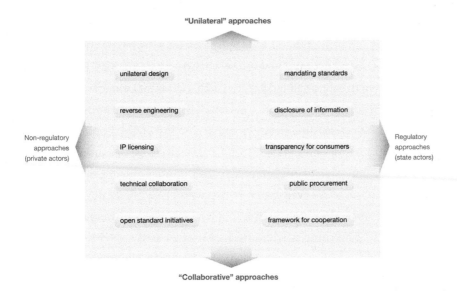

ability is to work within a single firm to interconnect the products that are offered to customers. For instance, Microsoft works hard to ensure that its Word and Excel programs integrate nicely with the Outlook e-mail program and PowerPoint. More often than not, interoperability is accomplished through collaboration between or among two or more firms. Microsoft, for example, has invested heavily in work with Novell to make the two firms' corporate technologies work better together than they used to. In the information business in particular, interoperability is often produced through standards processes, in which interested parties agree to definitions or requirements. They include a wide variety of approaches, ranging from "open" standards processes—that is, open, formal processes administered by standards organizations—to ad hoc cooperation. Office documents are rendered compatible over multiple types of systems, such as Word and OpenOffice, because most firms adhere to open standards for document formats.

In several cases that we have studied in depth—for instance, the business of health care records and the smart grid—market forces alone have not (yet) led to the level of interoperability that is desirable from a public policy perspective. In such instances, governments play a key role in fostering

interoperability. States have a broad range of tools available that help establish or maintain interoperability. Some of these instruments are more invasive than others. Government-imposed standards are a radical form of state
intervention. In safety-related areas—such as national security, emergency
communications, or navigation—society has an interest in establishing and
ensuring interoperability instantly and across the board. For instance, the
government mandates that those who sail or drive boats must use certain
modes and language to communicate with one another. In other sectors,
such as health care or energy provision, government can deploy other tools,
incentives, and "softer" approaches. For instance, the state might use its procurement or convening power to induce market actors, such as those who
sell health information systems, to aim for higher levels of interoperability.

The state will always be involved to one degree or another simply by
virtue of its role in shaping a business environment, legal framework, and
regulatory system that can facilitate (or thwart) interoperability efforts
across sectors. Nonetheless, there are varying degrees of government involvement, and in turn of private-sector leadership, in the promotion of
interop that will make sense. One or more of these approaches used in
combination might work to achieve the most advantageous interoperability within a complex system. In most of the cases we have studied, blended
approaches—involving diverse actors and one or more approaches concur-
rently—were applied to increase interoperability. Standards-setting initiatives among private actors that have been facilitated by government agencies
are one such example.

I t is not enough to achieve interoperability for existing systems. From
the design process through implementation, the goal must be sustainable interoperability, to guarantee that the systems will continue to work
together. At the same time, it is important to ensure that interoperability
over time does not lead to *lock-in*, a situation in which existing forms of interoperability become so standardized that they hamper innovation.

The most informal approaches, such as ad hoc collaboration among
firms, are usually the quickest route to interoperability. This can be seen

in the context of the social web: your Facebook account can easily connect to your Twitter account, or you can move a document from Google docs into another part of the Internet cloud with ease, even though these systems were all created by different companies.

It is quite another matter to ensure that you will be able to do so five years from now. Think of all those pictures you've uploaded to Facebook, Shutterfly, Photobucket, Flickr, Picasa, or Kodak Gallery (once upon a time called Ofoto). How do you know you'll be able to get them out, a generation from now, to show your grandkids? How do you know you'll be able to download them? Can you be sure that the data formats will be the same so that you can still view all those photos? How do you know those businesses will even exist? This problem, as we will see, has huge consequences for libraries and for our system of preserving knowledge and information in general. Interoperability can be the solution to these problems over time, too, but only if it is done right.

Interop can serve both to promote innovation and to thwart it. The vexing problem of lock-in hovers at the core of most interop debates. If the system remains flexible in the right ways and at the right levels of the interop stack, then higher levels of interop tend to lead to continued innovation over time. But too much interop, or the wrong kinds of interop, can have the opposite effect, causing a highly interconnected system, such as the global system of air traffic control, to become locked in to the technology of a particular era. We will return many times to the vexing problem of lock-in throughout this book. The lock-in problem helps clarify interop theory as a whole: interop is certainly desirable, but not all the time and not to the highest possible degree in every case.

Our theory of interop establishes a framework but not a single prescription, leaving most of the specifics of how to bring interop about to be determined on a case-by-case basis. That can feel unsatisfying. But it is an essential truth: the most interesting interop problems relate to society's most complex and most fundamental systems. Their answers are never simple to come by, nor are they easy to implement. This characteristic of interop theory is a feature, not a bug. It is the power of interop both as a lens and as a design principle that is relevant to so many big, intractable, interesting problems. The price to be paid for striving for a universal principle at the level of theory is that such a theory is full of nuances when it comes to application and practice.

We, as societies, must take interop seriously as we hurtle into a future full of increasingly complex and interconnected systems. Interop does not simply help us understand and navigate an increasingly interconnected world; it is also the invisible force that has enabled many great innovations, ranging from transportation systems to the Internet, and it will enable many more. The role of interoperability will become even more important in the future. The responses to the biggest challenges we face as societies, whether climate change or the health care crisis, require the smart use of technologies that connect unimaginably broad sources of information and knowledge, people, organizations, and governments. A sound theory of interop—the art and science of working together—will help break down the barriers that separate us, without creating new problems as we develop the complex systems of the future.

PART I

The Theory
of Interop

CHAPTER ONE

The Technology and Data Layers

An audience is gathered. Hundreds of people are waiting to hear a speaker begin a much-anticipated lecture. The appointed hour has come and gone, but there is still a group of people next to the podium whispering over the speaker's laptop; his presentation does not seem to work. People are staring up at the blank screen on the wall of the auditorium. The speaker is sweating as the tech guys try to get his computer to speak to the projection system so the images will appear on the screen. Members of the audience fidget in their seats.

It turns out that the speaker brought a Mac. The tech guys were expecting a PC. The speaker did not come prepared with the special dongle that enables the translation between Macs and most projection systems. A guy in the corner is on his cell phone, calling around frantically to see if he can find someone in the building with the extra part. One of the techies has already tried another option—translating the presentation on the Mac into

PowerPoint to use on the in-house computer—but the effort failed. The image-heavy presentation in the Keynote program would not render properly on the PC. The murmuring in the audience gets louder as the minutes pass. The speaker finally decides to start talking without the crutch of his presentation, while the tech guys continue to shuffle around near the stage waiting for the dongle to appear.

All users want their computers to be able to work seamlessly together with their cameras and smartphones. Nothing drives consumers crazier in the digital economy than technology systems that will not work together properly. As a general rule, interoperability *sells*. The web page that will not display videos properly without a complicated plug-in is doomed to lose out to one that requires the web surfer to do no work to make the videos play. The word processing system that can translate almost any document is likely to trump the complicated open source version that requires much more work and explanation. The social network that is easiest to use on any mobile device has a major leg up in the marketplace. (Granted, there are exceptions to this rule: music files that work only on iTunes and books that work only on specific e-readers. These special circumstances raise issues of their own.)

And yet, as we construct systems that work together in seamless ways, we need to examine the costs of this interconnection. We need to recognize that when we enable systems to share data easily and write computer codes to manage the handoff, we are ceding control of information about ourselves. We need to acknowledge that we are giving rise to concerns about data security. And we risk creating homogenous systems. We do not want all systems to be the same; in fact, it is important to maintain and facilitate diversity in the marketplace. We simply want systems to work together when we want them to and to not work together when we do not.

The clearest place to see these basic truths about interoperability is at the technology and data layers. The benefits and costs of interoperability are most apparent when technologies work together so that the data they exchange prove useful at the other end of the transaction. As consumers, we respond favorably to highly interoperable systems at the technology layer. We consistently prefer systems that work together without our asking them to and that make our lives simpler in the process. The data layer, a close cousin of the technology layer, turns out to be just as important: we need the data to be interoperable across systems as well. It is not enough merely to pass zeros and ones from one system to another, in digital-speak. The data must in fact be readable and understandable. Without interoperability at the technology and data layers, interoperability at the higher layers in our model—the human and institutional layers—is often impossible. But the job of getting the basics of interoperability right, even at the fundamental technology and data layers, can be deceptively hard.

I n today's digital economy, most of us use at least four types of basic information systems during a given week: we send and receive e-mails; we write, edit, and store documents; we send and receive data and voice signals over a mobile device; and we listen to music. For all these functions, we have lots of products to choose from. We can choose to use products and services made by Microsoft, Google, or Apple, among many other options. It is often easiest to stick with one company's products where possible. We also have the option to mix and match products and services from different firms, but that takes a bit more skill.

Corporate executives, focused primarily on functionality and reliability, often favor the suite of products offered by Microsoft. Corporate types can perform all four of the standard functions without having to use anyone else's products. Many big companies run all their e-mail through a Microsoft Exchange server,[1] perhaps even a server that is hosted in the cloud by Microsoft itself. (Most employees probably have no idea whether an Exchange server is running or not.) The employees use Microsoft Office to create and manage their documents (Word) and spreadsheets (Excel) and to handle their e-mail, calendar, and contacts (Outlook). Employees are given smartphones that run Microsoft's mobile operating system,[2] which syncs neatly with the Exchange server at their work. E-mail, calendar, and contact lists are all up-to-date at all times. These Microsoft customers listen to music on Microsoft's Zune. The experience in this world of information is likely to be completely seamless. Everything connects to everything else. It is easy to take for granted the seamlessness and the portability of information within this world of interoperability. Microsoft's engineers and designers have made all these products and services work together beautifully. The magic at work is vertical interoperability, or interoperability within the offerings of a single firm.

Those who are a bit more techie might favor instead a complete suite of Google's products. These people choose to manage their e-mail through the popular Gmail service and forward any other mail received into their Gmail account. For documents, they rely on the online Google docs service. They do not store anything on their computers locally; it is all saved in the cloud. They use HTC phones running on Google's Android platform because they love the many free apps on offer. Given their proclivity for openness over proprietary systems, they store all their music on their smartphone as MP3 files without any digital rights management software. As Google adds new services to its web-based empire, these users add them to their own computing world. The products work fast, and they work well together. The fact that nothing is stored locally does not bother Google's customers; they like to travel light.

And then there are the hipsters, the Apple users—a growing cohort, as it turns out, because, as of 2012, Apple is the largest technology company in the world. The hipsters are all Apple, all the time. They favor Apple's Mail program for e-mail and manage their information lives mostly through their iPads, synced regularly to their laptops and their iPhones. Their documents come in a variety of formats, including the version of Microsoft Word for the MacBook, which works perfectly well. They are devoted iPhone users. They were among the first to buy the original version of the iPhone and iPad, and they have upgraded each device, maybe even more than once, since. They love the touch screen and visit Apple's App Store regularly. Their music is all downloaded from the iTunes music store onto their laptops and then shared out to their iPhones and iPads. They occasionally have to go outside the Apple fold for their information-related needs, but it is rare. They benefit greatly from the connectivity and simplicity of the Apple product suite. These hipsters are the customers for whom the Apple team designs their intuitive, highly interoperable world.

In each of these three examples, the technologies involved are designed and implemented by the same firm. As Apple's iconic founder Steve Jobs was fond of saying, it was his intention to keep customers happy using only Apple products all the time. Customers are able to opt out, to choose a more complex experience using other firm's products, but Apple's systems are designed to keep them from straying from the true path. Google and Microsoft, likewise, do a very good job of providing a suite of services and applications that work together to meet customer needs.

This vertical integration within a single firm is the simplest, most common form of interoperability at the technology and data layers. A single firm makes a business decision: to develop systems internally that interoperate with one another. The experience for the consumer is positive: they are rewarded, through excellent functionality, for their loyalty. Firms have a strong incentive to make their own systems interoperate in precisely this fashion. If they do not, customers are liable to leave them in droves.

We take for granted much of the interoperability that makes our personal experience with information technology more seamless. We get annoyed when things do not work perfectly. Microsoft, Google, Apple, and their peer companies know this, of course, and strive for seamless interoperability within their product suites, but a great deal of cost and work goes into making it possible.

Although vertically oriented interoperability ordinarily makes for the most satisfying experience for the consumer, perfect integration among a single firm's products is the exception rather than the rule. It is far more common that people find a way to put a series of different services together to meet their needs. Today's computer user is more likely to run some combination of systems from these three companies or some other comparable mix.

There are many reasons to maintain the diversity of the applications we use. One is cost. For example, at least historically, Apple's appealing products have tended to cost more than the corresponding offerings of its competitors. As a result, even those who want to be all Apple, all the time, may not be able to afford such brand loyalty and tight integration. Another is the problem of legacy systems. Many users have been using BlackBerries, to take one case, for years and have their data stored in the formats that work well on other devices from Research in Motion (RIM). Technical decisions made by employers also contribute to diversity: people do not control the choices that their information technology (IT) departments make about what to deploy as an institutional e-mail system, operating system, or office productivity suite. Most of us end up running a hybrid series of applications that are offered by a range of companies.

The good news is that there is often a great deal of interoperability even between and among the services of different companies. It is not uncommon for an office worker today to run both PC and Mac operating systems, even on the same computer. More frequently, a worker might use primarily Microsoft systems at work, with a desktop PC (say, from Dell or Hewlett-Packard) that runs Outlook, Word, and Excel, with Exchange operating in the background to manage e-mail and other personal data. But the same worker may also use her Apple iPhone as her cell phone and to manage her music. She runs an Apple MacBook Pro for personal computing and an iPad for reading newspapers and books in digital form. For personal e-mail, she uses Gmail, and she uploads and shares documents using Google docs, both for work and at home. Other personal information, such as photos in Picasa and videos on YouTube and friends on Google+, is stored in Google's cloud-based services, easily at hand when she has logged on to Gmail to view her mail and her documents. This jumble of applications is not quite as elegant as the vertically integrated suite of products made by a single firm. But somehow, the pile of it works together pretty well.

Consider our worker's experience on vacation. At a picturesque spot along the shore, she asks a stranger to use her iPhone to take a photograph of her family. The image of her smiling family in the sunshine is stored locally on the iPhone. As soon as the stranger hands the phone back, she opens up her e-mail application, writes a quick note to her work friends to say "Wish you were here," and attaches the photograph. She hits send on the first message, then opens up a new e-mail and sends the same thing to her mother with the note "Check out your grandkids xoxoxoxo." She opens up an application on the phone and uploads the photograph to her Google+ page, tagging her husband in the process so that the photograph will show up on his page, too.

On the other end of the communication, thousands of miles away, her work colleague is finishing up her nine-to-five day in a cubicle buried within an office park; she sighs and smiles when she sees the photograph appear on the screen of her PC. The computer running the Microsoft products that the company deploys across the board does a perfect job of showing the picture, more or less immediately after it was sent from the seashore. The vacationer's mom, too, is able to view the picture without incident from the safety of her AOL account on her home computer. Other friends and relatives see the picture show up on Google+.

Replies flood back in to the vacationer's iPhone: two e-mails back, from the work colleague and from the mom, and notifications of Google+ comments, all of which can be read from the iPhone. When she gets back home, she will upload the picture to her Picasa album online for safekeeping over the long term. On a busy workday a year later, she might log on to Google, click on the Picasa link, and glance at the picture to enjoy the fond memory.

This vignette is unremarkable from the consumer's perspective. But a lot of computing magic has gone into making it possible. Dozens of invisible links make this apparent seamlessness possible. Sometimes computer engineers and designers are working directly together to ensure that this story can play out in a satisfactory way for the vacationer. But it is more likely that most of these interactions among systems are developed at arms' length, with engineers from different firms rarely if ever talking to one another.

Many things had to have happened to make the Apple, Microsoft, and Google pieces of this puzzle fit together without incident. The three firms had to have agreed—whether explicitly or, more probably, implicitly, without ever discussing it—to adopt a series of common formats, in this case for e-mail and for image files, that the vacationer used without knowing it. Apple needed to have developed its phone in such a way that the built-in camera recorded the picture in a format that would later be recognized by the Microsoft program that displayed the picture to the work colleague, the AOL program that showed it to the mom, and the Google+ system and the Google program that recorded it for long-term storage. Each of the firms also needed to have created an e-mail program that used the standard protocols for sending and receiving electronic mail. Each of the firms had to have connected their services to the freely available public Internet, using protocols for passing bits among devices over the network. And so forth. These invisible links among the programs are the magic behind interoperability.

The web is a great equalizer for technology firms. As consumers, we have come to expect that everything will work together without incident or interruption. We think it bizarre when something in the digitally networked world does not mesh with something else, perceiving whatever it is to be broken, in need of repair. This high degree of expectation is a powerful driver of interoperability. Market players are increasingly responding to this consumer demand and making these invisible links work for their customers without any government intervention.

Much of the time, today's big computing firms see making their services and products interoperate with those of their competitors as in their best interests because this interconnectivity serves their short-term interests. A smartphone that does not take photographs in a common format, enable a user to send those photographs immediately using a common e-mail protocol, or allow for instantaneous upload to Facebook and Twitter is not going to sell very well. A corporation is not going to install an e-mail system that cannot accommodate messages coming from outside its local area network. Office productivity software is not going to be widely adopted if it cannot work with legacy documents and spreadsheets or files that come across the network from the purchaser's clients, accountants, and lawyers.

For these reasons, some degree of interoperability is necessary if a product is to thrive in the information technology marketplace. The emergence of the social web demonstrates this imperative even more clearly. As we access more and more social web applications from our mobile devices, interoperability becomes increasingly useful. The fast-growing company Twitter, for instance, has made its system highly interoperable with virtually every kind of mobile device and operating system. On any new smartphone, using any major operating system, a Twitter user can download a simple application that interoperates seamlessly with a wide range of other services to let users easily post short messages or images from the smartphone to the Internet.

The mobile and social web is full of a rapidly growing set of examples, like Twitter and its related applications, of interoperability at the technology and data layers of everyday life. A great deal of innovation in the social web environment is taking place in location-based mobile services like Facebook, Twitter, foursquare, and SCVNGR. These services are applications that can be downloaded to mobile devices. They know where their customers are physically located, thanks to the wonders of geolocation. And they let them connect virtually with their friends or arrange to meet physically with those nearby.

The next generation of these tools, such as those created by the Dutch company Layar, create augmented realities that we can view through our smartphones.[3] These applications establish another layer of the web that places virtual information on real-space images. A person walking down a street in Manhattan could look through a mobile device at the front of a restaurant and be able to identify friends who are inside, see the menu, and receive a personalized bar-code-based coupon for a special price on drinks good for the next half hour. Or imagine how history could come alive through such a system. One might be able to see the image of the same street in 1750, in 1800, in 1850, in 1900, in 1950, and in 2000 through the screen of the smartphone. For important public spaces—think of Red Square in Moscow or Times Square in New York or Speaker's Corner in London's Hyde Park—the smartphone might bring up a video clip of something that happened at the same location at an important moment in world history. The social web—those applications that mediate our social lives online, such as Facebook, Twitter, and location-based services—forms an emerging frontier where these invisible links are at work more than ever before. The possibilities for the economy, education, innovation, and activism are invigorating.

Although interoperability at the technology and data layers is a great boon to society, there are also many ways in which it can pose real problems. These downsides fall into two general clusters: problems with getting to interoperability, and problems to which certain levels of interoperability can give rise.

Getting to interoperability in technology systems is not always quite so simple and harmonious as we have made it sound so far. It's not always the case that, thanks solely to the magic of the market at work, firms decide to make their systems interoperate for the benefit of their customers in all ways at all times. The interests of businesses and their customers are not always perfectly aligned. Companies want customers to favor their products and services; customers are more likely to want to get a particular task done. Companies may want to make it more difficult to move personal information from one place to another to make it hard to switch services, which customers may want to do for any number of reasons. Social networks are such an example: it is easy to connect your Facebook account to the newest web service, but it is impossible to move your entire social graph—all your connections—to the new service.

Sometimes firms create interoperability that costs a lot of money to pull off and, in turn, pass those costs directly on to the consumer. Consider what happens to a business traveler who needs to get to Europe on short notice. His trip is quickly booked and he manages, without incident, to get himself physically to Munich in time for his meetings. He communicates with his spouse back home, checks Facebook from his smartphone, and downloads his e-mail—all on the network of an international partner of his home service provider. His next cell phone bill, a month later, shows charges of $487.34 for "roaming data services" and a couple of expensive international calls. In this case, the two mobile service providers have figured out how to give travelers a seamless communications experience, but they charge a hefty fee for that privilege. There is nothing wrong with the market working this way, so long as travelers know that they will be hit with such high charges in order to receive this kind of interoperability. Sometimes firms offer high levels of interoperability for free; when the market will bear it, they can charge substantial amounts for providing a similar level of interoperability. Travelers can get voice and data connectivity even on another company's system, but it may come at a very high cost. There is no law that says that all systems must interoperate globally and seamlessly for free.

In other, more troubling cases, firms decide not to allow high levels of interoperability at all. Amazon's Kindle is a case in point. The Kindle began as a largely closed system and has grown steadily more interoperable over time. Customers buy the Kindle device of their choice. (As of December 2010, Kindle was the company's biggest-selling item.)[4] When they get it, they log back on to Amazon's website and begin buying and downloading books. Suppose, over time, a customer downloads a thousand books. With great delight, she reads half the books she downloaded. But then she leaves the Kindle device on a train on her way home one night, and it never makes its way to the lost and found.

Here is where the first big interoperability problem arises. Although Amazon enables the Kindle to interoperate in some respects with the services of other firms, such as the daily download of the *New York Times* or the RSS feeds of bloggers, it has not made the Kindle work with the formats of other e-book providers. The customer faces a tough choice. She can buy another Kindle so that she can access and read the thousand books she has already bought, and she can continue with the Kindle platform indefinitely. But if she does not stick with Amazon and its services, she is out of luck. She cannot buy a Sony Reader or a Barnes & Noble Nook and simply transfer her books over to the new platform. (She can, however, read her books on an iPad, if she wants to fork over the money for one. Today, Kindle can also run on Android-powered devices, too.) The same goes for the customers of the other e-book services, by and large. There is very little portability of electronic book content from platform to platform. These systems have not been made to interoperate horizontally. Companies are employing non-interoperable, proprietary standards in a way that frustrates consumers.

The same problem arises with digital music. Users of Apple's iTunes who buy the inexpensive versions of the songs they download may someday face the same choice as Kindle buyers. If they stick with the vertically integrated Apple services and continue to buy new iPods when the old ones break or get lost, everything is fine. As soon as they want to move all their songs over to another format or to use a device offered by another company, at least in the original model of iTunes, they have a problem. With iTunes, increasingly, customers do have a choice: they can pay somewhat more per song in order to get files that are not encoded to play only on iTunes. Apple has done a very smart thing: it began using differential pricing, or price discrimination, to enable data portability at a cost. More recently, it has gone even further to make its songs available in such a way that they can be moved to another service.

Pause here to consider the difference between the old analog world and the new, digital-plus era in which we live. Previously, consumers bought books and records the old-fashioned way. They went to stores, paid cash for physical goods, took them home, put them on shelves, and did what they wanted with them. Today, it is quite different. They pay online, using a credit card or a PayPal account, to get access to virtual goods.

But customers do not have all the same rights that they used to have with respect to those goods. They cannot move them around in quite the same way. And sometimes, depending on their pricing plans, they don't own the goods at all; they are just leasing them. When they stop paying for a monthly music service, they lose access to all that music. They are also bound by a contract for what they can do even with the books and music they have "bought." Instead of having the right to resell their books, records, or CDs, which is permitted under the law's "first sale doctrine," they are bound by a contract with a service provider forbidding them to resell digital goods. This absence of a digital first sale doctrine will come to be a major problem for consumers, especially readers and music lovers. We should not be doomed to lease our cultural heritage only to lose what we have paid for if the lessor goes out of business or if we stop paying the cost of the lease.

For the time being, consumers still have the choice of buying physical goods as well as virtual goods. They can opt out of the digital economy in this and many other ways, making this era a "digital-plus" rather than a purely digital one. Materials are produced in a born-digital format: a book, such as this one, begins as a digital good (a Word document, in fact) that is then rendered in various formats, including as an e-book and as a hardback (and, with luck, as a paperback, if enough people buy the hardback). But it is becoming increasingly difficult to opt out of the digital economy altogether. The importance of digital interoperability grows as our collective dependence on digital goods and services grows.

In addition to opting out of the digital economy, as consumers we have yet another choice, which is to create our own digital environments independent of the big commercial services. Thanks to open source software development communities, those with reasonably strong technical skills and a bit of patience can create an alternative environment to the Google /Microsoft/Apple-only worlds that most of us inhabit. They can run a computer with an open source operating system, run open source e-mail and productivity tools, use an Android phone and put lots of open source applications onto it, and use broadly available open music systems to enjoy digital music.

Even the most technically skilled users will find roadblocks associated with the roll-your-own approach, though. If they e-mail basic text documents in the Open Document Format (with a filename extension of .odt) to friends using a standard form of Microsoft Word, the recipients may have a tricky time making the conversion work. It is almost certainly possible to download some kind of software patch that will make the system interoperate, but that is likely to cause frustration and lessen productivity, in terms of speed at a minimum.

And some systems simply will not play nicely with such an all-open-source approach. Unless you are willing to turn into a hacker and take a good deal of legal risk, you will not be able to access some kinds of content—such as e-books and digital music—that is available only using certain forms of technical protections. If you want to use something more exotic than the

Google/Apple/Microsoft-style services, you have that option, and you may even get some better results for your efforts. It is crucial that this do-it-yourself option exist. But you will also run into some hitches when you try to interoperate with other people and some kinds of content—not because the technologies cannot eventually be made to work with a little ingenuity, but because it requires extra work and knowledge at the human layer of interoperability, the subject of Chapter 2.[5]

The second big problem associated with interoperability is the increasingly urgent issue of personal data privacy. The e-reader example given above is a starting point. In the analog world and before the introduction of digital frequent-buyer cards, when a reader bought a book at a physical bookstore, there was no way to track that book over time. There was no way, short of tapping the clerk's memory of who bought what, to determine what customer bought a given title. The person who bought the book, moreover, could easily sell or give the book to a neighbor in the historically vibrant secondary market for books. Every once in a while, someone in law enforcement authority might try to learn who bought what book as part of an investigation, but the process was much less likely to produce an answer in the analog world.

In our interoperable and digital world, the system ensures that Amazon knows exactly who bought which book from it. This kind of tracking has some upsides for the consumer: Amazon, for instance, can recommend other titles customers might like based on their previous purchases, just as TiVo will recommend new shows based on what its technology determines to be a viewer's taste. Some customers today enjoy sharing their purchase histories with their friends and tracking the purchase histories of others, on Amazon or other services. Certainly, this kind of data collection benefits the companies selling the product: as companies come to know their customer bases more accurately, they can make better decisions about what products to produce, how to market them, how to price them, and so forth, giving rise to increased margins. That improved efficiency might redound to the benefit not just of the publishers and the booksellers but also of the authors (through fatter royalty checks) and society at large (through lower prices and transaction costs). The cost of this heightened interconnectedness at the technology and data layers—in the case of Kindle, among the book producers, booksellers, payment systems, and customers—is substantial for consumers when we see it as connected to concerns about our privacy.

The context for this growing fear about personal data privacy extends beyond the world of e-books, of course. We are giving up control of information about ourselves, not just in the context of book sales but in the context of nearly every form of social, commercial, and political interaction of our lives. When we download a new mobile application for our smartphone, connect with our friends in a geolocation-powered social network by checking in to a particular venue, or give money to a political candidate, these applications swap data about us across technology platforms without asking permission. When they do ask permission, it is buried within an extraordinarily long legal document—the terms of service or privacy policy—that virtually no consumer ever reads in detail. By the way, those "details" usually reserve to the company the right to share the information in most instances with their partners, and we as consumers "agree" to this arrangement when we click-through to use the service.

Increasingly, the party we fear is not the jackbooted law enforcement officer or even the companies we know; rather, it is the unknown corporate players who use strategies based on interoperability in order to aggregate enormous amounts of data about us without our knowledge. In a world of data aggregators, such as Spokeo and ChoicePoint, to name two, we are not wrong to have these fears.[6] Increasingly, companies (and political campaigns, for that matter) have reason to keep these data about us for long periods of time, to be able to target us better with new offers. The primary mechanism for making money from digital services that appear to be "free" to consumers is to target advertisements of various sorts to us. That targeting grows more effective the more these companies know about us. As we allow more and more services to interoperate, we set up the possibility of more and more data sharing. There are ways to set limits—through contracts,

technology, and federal laws, for instance—but we will have to do so deliberately. Companies on their own have not shown much restraint in this
respect during the first few decades of the digital world.

We ask for these systems to work together seamlessly, of course, but we
probably do not want the consequences of unfettered interconnection. This
idea helps illuminate the distinction between the technology layer and the
data layer: we may well wish for the technologies to be able to work together,
but we may not want all the data from one service to flow into the other
without any constraints. Just because we can use our Facebook account to
log on to our account on a new social application, we do not want those two
parties to share all their data with one another automatically.

The addition of the mobile web to the story makes the privacy concerns
come to life even more vividly. By linking systems at the technology and
data layers, and by overlaying the use of geolocation and powerful mapping
applications, we can tie people, places, and times together. This creates
enormous potential benefits for learning. Think of the advantages to
schoolchildren on a field trip as they make their way down the Mall in
Washington, D.C., able to call up all manner of perfectly related, detailed,
rich teaching materials. High-tech scavenger hunts already enliven school
field trips as the cutting edge of applying technology.

But that same connectivity may concern us if those schoolchildren are
tracked, as a group or individually, based on their use of the tools, mapped
to time, place, and activity. And if we do not watch out for it, we may become too complacent about the extent to which interesting information of
this kind is fed to us through highly connected devices rather than being
something we seek out. As we design these wonderful new technology systems, we need to ensure that the human qualities that we care about—like
personal privacy and active learning—are at least preserved, if not enhanced, in this new world.

The dystopic view of the highly interoperable future—in which technologies allow companies, unchecked, to share information about our reading, spending, and walking patterns—does not need to come to pass. But
the risk is very real, and it is related to the degree and type of interoperabil-

ity that we establish between and among technologies. As we construct an increasingly interconnected world at the technology and data layers, we need to ensure that the system does not come with costs in other areas— such as privacy and security—that are higher than we are willing to pay. A theory of interoperability by design that builds in privacy and security protections from the start can help enormously in this respect.

The technology and data layers make up the foundations of our interop theory. Interoperability often only becomes effective, though, when we are able to establish more subtle and complex forms of interop. We characterize these additional layers of interop as the human and institutional layers. The human dimension of whether and how people can understand one another and work together often matters far more than whether technologies can pass data back and forth. And last, the cultural issues associated with institutions of many sorts, whether organizations or entire legal systems, are often the hardest forms of interoperability to recognize. They can sometimes also be the most important to accomplish.

CHAPTER TWO

The Human and Institutional Layers

In December 1903, Wilbur and Orville Wright embarked on the world's first controlled airplane flight. Their short trip revolutionized the world of transportation forever. More than a century later, air travel is a multi-billion-dollar business on which virtually everyone in the modern economy relies one way or another. Growth continues with each passing decade: in 2010, 786.7 million passengers traveled on 10 million flights coming through the United States alone.[1] Asian Pacific carriers, meanwhile, are growing their seat capacity by as much as 10 percent per year.

On that first day, the Wright brothers had the airspace they needed for their trip all to themselves. Now the skies are crowded with planes. The need for air traffic systems to coordinate these flights has grown steadily over the past century. As more flights take off each day, the complexity of the problem grows, as does the importance of getting the system right. The creation of reliable and interoperable air traffic systems has become a priority for governments and the aviation industry around the globe.

Interop is a key to safe, well-coordinated air travel. It can save lives and improve economies. The air traffic example helps break down the problem of interoperability yet further, into its four layers. The technology and data layers are foundational. It is crucial that pilots, for instance, have the technology to communicate with other pilots and with air traffic controllers on the ground. They need to be able to share data about their planes' locations, speeds, and trajectories in order to facilitate coordination and avoid collision. But with an international system such as aviation, the human and institutional layers are also essential. It is not enough for the technology involved to share the information from pilot to pilot or from pilot to air traffic controller. This complex system must work in a diverse global environment in which the language, locale, and cultural norms on the ground below routinely shift over the course of a multihour flight. It must enable competing commercial actors to work with one another and with government agencies in high-pressure situations. And it must be able to adapt when conditions change.

The importance of the human and institutional layers in aviation is, alas, best illustrated by systemic failures. Consider the 1977 Tenerife disaster, in which two Boeing 747 passenger aircraft collided on the runway of Los Rodeos Airport. This devastating crash resulted in 583 deaths, the deadliest accident in aviation history. Although there were multiple contributing factors, including bad weather conditions, language problems featured heavily in the accident. An analysis of the cockpit voice recorder by crash investigators revealed that the captain of one aircraft was convinced that he had been cleared for takeoff, whereas the control tower was certain that the aircraft was stationary at the end of the runway and awaiting takeoff clearance. The use of ambiguous phrases contributed to the problem. One copilot said, "We're at takeoff." Controllers in the Tenerife control tower responded, "OK," which the captain interpreted as clearance for takeoff.

The Tenerife disaster demonstrates why interoperability at the human and institutional layers is just as essential as interoperability at the technology and data layers. The problem at Tenerife was not that the copilot and the controllers could not exchange data with one another; the technology worked fine, and the data flowed right through. The problem occurred at the human and institutional layers. The organizations involved had not sufficiently coordinated the language they would use in such a circumstance. The system of air traffic control, in this case, did not prevent the use of ambiguous language, which led to a tragic disaster. Most stories of interoperability, fortunately, do not involve life and death. But there are plenty of examples of people and institutions failing to work together adequately.

Interop at the human and institutional levels can lead to problems as well as help solve them. One of the major pitfalls of interop, one easily seen through the lens of the human and institutional levels, is the problem of lock-in. The achievement of interop itself can hinder the development of new, better systems of keeping people safe and creating jobs. Systems at high levels of interoperability can get locked in to a specific set of technologies and protocols that may work well on day one but become suboptimal over time. The effect of lock-in is ordinarily to preclude necessary innovation. As we put in place interoperable systems, we need to design and implement them in ways that will ensure that change can occur over time. This need for flexibility and innovation can be seen through a close examination of the human and institutional layers of interoperability.

The human and institutional levels of interop tend to build on top of the technology and data layers. In the case of air traffic control, however, the first interoperability was accomplished at the human and institutional levels. This type of interoperability took the form of shared rules and signals. Later, technological and data interoperability came into play as telecommunications equipment became more sophisticated and useful. The interoperability at these four layers—technology, data, human, and institutional—established a complex, highly interconnected system that has saved a lot of lives. It has also proved very hard to change to accommodate innovations. The problem of lock-in, when systems get highly integrated at the human and institutional levels in particular, is a major concern about interop in the most complex systems.

The early interoperability in the air traffic control system had its focus at the institutional layer. In the golden years of aviation—between World War I and World War II—a series of basic air traffic rules emerged. These rules covered aircraft identification, safety standards, and common navigation practices. Rules make up a central element of what economists and legal scholars refer to as institutions, one of the reasons we refer to this layer as institutional. Europe was the early leader in developing an air traffic system that included interoperable guidelines and procedures. In 1922 the International Commission for Air Navigation developed general rules for air traffic that were applied in most countries. (The United States, back then, was an exception: a separate set of air traffic standards was created under the auspices of the US Department of Commerce.)

The air traffic rules enacted by local governments and international organizations in the early days of aviation were important steps forward in improving flight safety. These standards were not much more than a basic set of traffic rules for pilots. One such rule: "Don't begin take-off until preceding aircraft are clear of the field."[2] The rules were simple and clear, and they were a starting point for the system of air traffic management.

It soon became clear that these basic rules—this institutional approach to interoperability—would be insufficient to keep pilots and their passengers safe. Additional instruments and communication protocols were needed to manage the fast-growing air traffic and to avoid collisions. The development of air traffic *control* (ATC) systems in the late 1920s responded to this need. For starters, air traffic control necessitated the hiring of people to serve as "controllers" on the ground, by waving brightly colored flags at pilots in the air. The system required nothing more than a human being—a flagman—who shared an understanding with the pilots of the meanings of the different flags. (High visibility helped, too.) The flagmen and the system of signals, the meanings conveyed by the flags, demonstrate interop at the human layer in the early air traffic control system.

The system necessarily grew more complex as the number of participants increased. As air traffic grew rapidly, so too did the danger of collisions. This complexity gave rise to the integration of technological interoperability as well, which came to supplement the human and institutional forms of safety measures. The need for more-sophisticated communication protocols and related methods of coordination was clear. Flagmen were soon replaced by radio-equipped airport traffic control towers that enabled air traffic controllers to communicate directly with flights in ways that did not rely upon whether the day was clear and sunny or not. By the end of the 1920s, European airlines were required to equip all passenger planes with radio receivers. The United States followed suit a decade later. Radio-equipped control towers were established at airports across the country.

Engineers designed the radio system so that it was interoperable at the technology and data layers. It enabled two-way voice communication between pilots and air traffic controllers. It enabled improved navigation as well. By the mid-1930s, in order to handle the increasing number of flights, air traffic control was no longer limited to airport areas but had expanded to the respective airways to provide controls along the planes' routes, not just at the endpoints. In the years that followed, radar technology was added to the ATC system. It was installed first on the ground and, later on, inside the airplanes themselves. Radio spectra, aircraft instrument settings, and navigation checkpoints were standardized.

The first shift these systems had to respond to was the growth in air traffic from one plane (the scenario of the Wright brothers) to a modest number of planes carrying a modest number of passengers. The early air traffic control systems held up remarkably well in an age in which controllers' primary concern was to ensure that planes were kept distant from one another. But the problem has changed once again: as air traffic increases, it has become impossible to rely merely upon physical separation of the planes, in the air or as they seek to take off from and land at the same airstrips, to keep passengers and crew safe. Now air traffic control systems need to help pilots navigate in close proximity to other planes.

New technologies, such as Global Positioning Satellite (GPS) systems, could help meet this new and harder challenge for air traffic control. GPS has been around for quite a while. What is surprising about the air traffic control story is how hard it has been to integrate new, and plainly superior, technologies into the air traffic control systems globally. So far, it has been nearly impossible to replace the highly standardized legacy technology to create the next-generation ATC system.

The problem of updating the ATC system with GPS shows the importance of the human and institutional layers of the model and how they interact with the technological. The technology exists to improve upon the existing air traffic system. But we have not been able to integrate it effectively into the organizations that manage and rely upon the system, such as government agencies and the airlines. The problem is that the current system works well enough to avoid most problems. The institutions and people involved work together reasonably well using the existing system. The incentive to change is not yet high enough to overcome the inertia that develops in systems that are highly interoperable at the human and institutional layers. Most of us do not like change, and we avoid taking on hard tasks such as implementing complex new systems that are deeply embedded within institutions.

What has taken place in the ATC system is an instance of lock-in at the level of a good, but not great, set of technologies and protocols. The broad adoption of interoperable standards at the technology, data, human, and institutional layers has created a system that is locked in to a previous generation's technologies and processes. Lock-in to a given system tends to occur when the practices involved are deeply embedded in the human and institutional layers, as they are in the air traffic control example. Individuals, companies, and governments in hundreds of countries have come to rely on a series of rules, norms, technologies, and flows of data that reach a deep level of interoperability in the daily procedures of everyone involved in air traffic. This depth and level of consistency is terrific at the outset, but it can cause problems when changing those processes or technologies would help to improve the system. This problem of lock-in is one of the core puzzles of interoperability, to which we will return several times.

I nteroperability sounds like a technical concept. It evokes gears that interlock with one another or massive data flows across corporate firewalls. But it turns out that the human aspects of interoperability are often just as, and perhaps even more, important than the technological. The human element is always essential to understanding interoperability. How we work together as humans, often relying upon technological tools to communicate, can determine whether the most seamlessly interoperable technologies prove effective for their given task. In the context of air traffic control, as well as other public safety examples, language is often at the crux of human interoperability.

Effective, frictionless communications among humans is essential to air traffic safety. In the wake of Tenerife and other accidents, sweeping changes were made to international airline regulations. One set of changes involved the international standardization of airport codes, instrument calibration, and other technical issues.[3] The United Nations–sponsored International Civil Aviation Organization (ICAO) had long before set language-proficiency guidelines for air traffic control, including guidelines aiming to create higher levels of interoperability among the people involved in flight operations. In the late 1970s, the ICAO enacted guidelines that required all pilots of international fights to be fluent either in English or in the native language of the country they are flying to. It was only in 2008, though, that the ICAO mandated that all communications among members of the air traffic community be conducted in a single language (English).

The mandate of English as the new lingua franca of ATC is a necessary but not sufficient step toward human interoperability among pilots and controllers. In order to reduce miscommunication further, the ICAO has developed a standardized phraseology—a simplified version of English—that is now required in all situations for which it has been specified. Plain English is only to be used as a last resort in radio communications, only when the standardized phraseology cannot serve an intended transmission. ICAO also recognized that language proficiency is not only about words but, rather, is multifaceted and multilayered. Especially in emergency situations, the nuances of language—such as pronunciation or grammatical structure—matter too.

Language is the clearest way to demonstrate the need for interoperability at the human layer. It, too, is more complex than we have made it seem so far. In the context of complex systems, it is not enough to say that we "speak a common language." Those who study linguistics break the problem down further in ways that help us realize the depth and intricacy of interop problems among complex systems.

Think of the interop problem as having layers within the layers. Inside the human layer, we can see that people have to work together in order to make use of the interoperability at the technology layer (the radios operating on the same frequency, for instance) and the data layer (the information coming through from one radio to another). In order to make the communication work, there needs to be interoperability within those layers, as well.

Language is just one component of human interoperability, but even that is far from simple. Within language, one might break the problem down into semantic interoperability (that is, there must be agreement as to what type of information the language is conveying), syntactic interoperability (agreement on what part each unit of the language plays in the communication), and lexical interoperability (agreement on what specific terms mean).

When the information is being shared from one system to another, the systems must agree at the level of semantics: roughly, on what is meant by a series of signals conveyed through the system. If data come through from one system to another, for instance, the receiver needs to know whether those data correspond to the date when the information was created, a time stamp, or the message itself.

The syntactic level of interoperability ensures that the receiver is able to interpret the data in the right way. In the air traffic control context, information about when a given flight is scheduled to take off and land needs to be shared across several institutions: the airline, the airport, and government agencies all need to have this information. It is necessary but not sufficient for the receiver to know that the data represent a time, a date, or the message text. At the syntactic interoperability level, the system will still need to be able to interpret the data received. If the data correspond to a date and a time for a flight's expected takeoff from a given airport, the systems must have agreed in advance as to the syntax, as to what each part of the data conveys. By having an agreed-upon syntax, the receiver knows that the numbers at the start of the sequence correspond to the year, the middle numbers refer to the month, and the final numbers refer to the day. (Think of it from the from the traveler's perspective. The customs forms one fills out before disembarking in the United States require the date to appear *before* the month, contrary to the ordinary convention in this country.)

Once the semantics and the syntax have been worked out, then the air traffic controllers can focus on what the conveyed message means, or "lexical" interoperability. The message about the flight's departure time includes a given word to represent a month (*April*, in English, as opposed to *avril*, in French), the receiving system or person needs to know what that word means. Thanks to interoperability at the semantic level, they know that it is part of the message itself (and not data about the data, for instance); thanks to interoperability at the syntactic layer, they know that the letters are meant to convey the month in which a plane will be taking off; and at the lexical level, they will come to know that that specific combination of those five letters—*a, p, r, i, l*—corresponds to the fourth month in the calendar. The human beings involved in coordinating which flight takes off next can then go about their business, which is now conducted globally in English, in the interest of interoperability at the human layer.

There are many ways that human interoperability turns out to be highly important in complex systems. Effective communication through language is just one of those ways. People need to be willing to work together, and, as with any skill, they have to learn how and practice. Often, we need incentives to cooperate rather than focusing on our parochial interests. We often work together more effectively when we also spend time together in relaxed, social settings; we have more empathy and willingness to pick up slack from another person. When we ignore the human element of interoperability, we are likely to be setting up the system to fail.

The story of air traffic control demonstrates how a multi-billion-dollar industry, together with government agencies, have teamed up to invest heavy resources to achieve higher levels of interoperability at all layers. But there are even more ambitious—and farther-reaching—interop initiatives underway around the world. As these systems grow more complex, so, too, do the interop-related problems grow more complex, in terms of both how to accomplish interop and how to mitigate its downsides. Interoperability strategies, as opposed to strategies that make everything the same, can take a great deal of time to implement, and the variability in the ways they are implemented can introduce cost and complexity.

The process of creating the European Union (EU) demonstrates the benefits and the costs of seeking high levels of interop at the human and institutional levels. People from vastly different cultures, with diverse histories, have jointly created a common system of governance, trade, and currency. Despite their many differences, they have developed a political union, a common currency, and ways for organizations and people to work more effectively together. The drivers are many: global competitiveness, the desire for political power as a region, reduced barriers to trade, increased coordination within the region, and so forth. But the people of Europe do not want to go so far as the highest level of interconnection, to become a single state. They want to preserve local differences—diversity— and the innovation that comes from heterogeneity. Striking this sensitive balance is a tricky business.

In the long process of creating the European Union, the challenges of interoperability have played out at many levels. Getting multiple nations to work together is more complicated even than making a few software systems mesh. Although interop can mean putting together a system of interoperable software and data standards, it can also mean something much more complex: making people and institutions, as well as the systems on which they rely, work together. And even when the systems are made interoperable in these ways, problems can still stem from the complexity of the system established.

At the institutional layer of interoperability, organizations, firms, and states jointly move toward common goals. To make things interoperable in complex systems, it is crucial to link up organizational structures, business processes, and groups of people in ways that enable them to work together effectively. The alignment of business structures and processes that human beings follow turns out to be essential in achieving higher levels of interoperability in the example of the European Union. This is the distinction between the human and institutional layers of interoperability: human interoperability allows individuals to work together—for instance, through effective communication across networks—whereas institutional interoperability ensures that groups and entire states are able to work together optimally.

There is an essential difference between making a complex system interoperable and simply making everything the same. Europe is a perfect example of this core tenet of interoperability. There is no chance that everyone in Europe is going to agree to speak the same language (remember Esperanto?). Nor is it plausible that every state in Europe will agree that every local law shall be made by a single body. At the same time, each state participates in the European Union in order to cooperate more effectively and share in the benefits of a higher level of interop than they would have without the coordination functions that the Union makes possible.

Language, once again, proves instructive. Language is the most basic interop problem facing the European Union. Among the member states, there are twenty-three official languages. At the most basic, human layer, people across the region need to be able to speak to one another. This problem, of course, is not new; since time immemorial, the people of different cultures have had to either learn one another's languages or hire translators to ensure that communication works across geographic and cultural boundaries.

In Europe today, the issue of language interoperability is central to the proper functioning of markets, legal systems, learning, and cultural exchange. The way the European Union has sought to solve the human interoperability problem is not to enforce a single language but, rather, to establish a system of linguistic interoperability. In certain contexts, people agree to speak a certain language to one another for simplicity. In other contexts, rules require that documents be translated and published in the twenty-three languages that currently define the European Union. This priority issue is even enshrined in government, with a European Commissioner for Multilingualism.

Language interoperability operates at the human layer. Legal interoperability demonstrates the same concept at the institutional layer. In the European Union, the goal has not been to create precisely the same legal system in every member state but, rather, to establish a mechanism that has enough common ground to work, while honoring local diversity. The European solution has been a system built on directives, which has proven remarkably effective over the years, given the challenging setup of making things work across the broad region of Europe.

Here is how it works. Representatives of the various states come together in the European capital of Brussels to work together on topics of common interest. The representatives ultimately agree to broad directives that must then be translated—or, as a lawyer would say, transposed—into the local law of each state. The idea is not that every law must end up being exactly the same; approaches to solving problems might be different, as might the specific implementations of the relevant laws. But the laws are all meant to interoperate at an optimal level. Across the European Union, information, people, and trade can flow more freely, thanks to these directives and the laws that derive from them, without all the laws across the region having to become the same.

An example of the directive system at work is the way Europe has addressed data privacy. The European Union has taken a strong stand on protections for personal data that are collected by institutions. There was no way, however, to create a single privacy law for all of Europe that would be sensitive to all local norms while also achieving collective goals. The EU Data Protection Directive was an effective solution. The European Union enacted a high-level rule that required each state to then translate into local law a series of principles regulating the use of data about individuals.

The problems of the directive approach, as elegant as it sounds in theory, are evident in every case of transposition. First, the process is very time intensive. States transpose at different speeds, and none of those speeds are set at "fast." In the case of the EU Data Protection Directive, the effects of the law have taken many years to reach the people it is intended to protect. And the protection is, by design, variable: not every state is transposing the directive as fully as it might or enforcing the rules as vigorously as others might. The net effect is that the system is not uniform across Europe. From a business perspective, this complexity comes with a cost. It is not enough to know that you comply with the directive; if you compete in multiple markets within Europe, you still need to comply with local laws with respect to data privacy as well as to ensure that you are in compliance with EU law. There is more bureaucracy involved, not less, from the perspective of businesses. These problems show the limits of an interoperability strategy: its flexibility and lack of uniformity come with costs in terms of time and complexity.

Institutional interoperability can also be seen in the way organizations are set up to work together. The European Commission recognized that it is not enough to ensure that the languages and the laws are interoperable. The governments of the member states need to work together to ensure that citizens and businesses can fully profit from the single market.

The problem is that government actors need to be able to work together to recognize citizens and organizations from other European states in order to render them services that previously were reserved for their own citizens. For instance, within European airports, the passports of European citizens are recognized with a special passport control system that allows them faster entry than available to those from outside Europe. The problem grows much more complex in the context of trade, currency, and regulations: systems that worked at a state level need to be adapted so that they function at a regional level, without eliminating all aspects of local diversity, which are meant to be maintained.

The commission has sought to solve this problem through an initiative designed to encourage public administrations across the EU to work together so that citizens and businesses can get information, provide documentation, or obtain permissions from the governments of other member states. The European Interoperability Framework[4] highlights three areas that public administrations need to address to be able to work together efficiently and effectively. First, they are asked to align their existing business processes or even define new processes and to make sure that those processes are well documented. Second, public administrations are urged to clarify their organizational relationships through such instruments as memoranda of understanding, service level agreements, or multilateral agreements. Third, the different government entities are asked to agree on a set of change management processes to ensure accuracy, reliability, and continuity of services despite the rapidly changing environment.

As states have gone about implementing the structures and processes of the European Union, they have established ways for their institutions to work together. The effect has not been to make, for example, a single, European, ministry of trade; the notion has been to adapt each state's ministry of trade to be able to meet local demands while ensuring that they can all work together effectively on common goals and with the commission's bureaucracy in Brussels. Nor has every privacy law in Europe become identical. But information, currency, people, and trade can—at least in theory—flow more freely across borders in the European Union than ever before, with a corresponding series of benefits to each participant.

The hardest interoperability problems often have nothing to do with making technologies work together or enabling data to flow across systems. The hardest problems often arise in linking together entire business processes and workflows across otherwise uncoordinated people and organizations. To create meaningful interoperability across institutions, leaders have to create the expectation and incentive for cooperation among units and entities. A successful approach to interop means not only addressing the technological barriers to interoperability, but also overcoming human roadblocks by educating, training, and assisting all team members to develop the culture, awareness, and skills needed to achieve and maintain interoperability. An effective interoperability strategy needs to take into account the potential at each of the four layers in the model. And once interop is accomplished in a complex system in these interconnected ways, the system will still need to be flexible, to be able to change over time to accommodate innovation—no mean feat.

This is what is special about interoperability: it is not about everything becoming the same; rather, it means figuring out what aspects of a complex system must work together in order to accomplish shared goals. Very often, the most important parts needing to be rendered interoperable are humans and institutions, not technologies and data. Most of the time, we need to find optimal levels of interoperability at all four layers: technological, data, human, and institutional. The promise is that we can accomplish the benefits both of diversity and of collaboration in our increasingly interconnected, global society.

PART II

The Promise and Perils of Interop

CHAPTER THREE

Consumer Empowerment

C ustomers tend to want more, not less, interoperability, a demand that is a major driver of interoperability. For ordinary consumers, interop means that smartphones can work seamlessly with laptops and stereo systems, all sharing the same music files, purchased just once from an online service. In turn, customers have the ability to do more with the information systems they use. Interoperability provides consumers greater choice and autonomy.

The benefits of interop extend beyond the individual consumer and deeply into the business world. Demand from business customers has also been a major driver of interoperability in the information technology sector. Interop empowers corporate customers to make systems work together behind the scenes: for instance, to make Microsoft products work with Novell's to create a more effective corporate IT system, on which office workers can accomplish more.

Interoperability tends to be taken for granted, so its value is often easiest to see in situations where things do not work together nicely. Take, for example, battery chargers for electronic devices. As researchers in the field of digital technologies, we are obsessed with all sorts of gadgets—cell phones, laptops, tablet computers, portable music players, and e-book readers. As a result of this obsession (we call it professional interest, especially when challenged by family members to justify our credit card bills), we carry an increasing number of consumer electronics products with us on every trip. Over the years, we have accumulated a large collection of chargers for our devices. With every new cell phone, for instance, comes a new charger with a different plug, often even when the same manufacturer made the charger and the cell phone. Every year, 51,000 tons of redundant cell phone chargers are produced. The lack of interoperability among device chargers is a big, expensive annoyance for consumers. (It is also terrible for the environment. According to one industry association, eliminating charger redundancy would reduce greenhouse gas emissions by 13.6 million tons annually.)[1]

Every consumer would prefer to have chargers that work with any smartphone, yet many still don't have that option. This is one of the conundrums of interop: it is often deceptively hard to achieve optimal levels of interop, even when it is plain that nearly everyone would benefit from greater cooperation. The cell phone charger example is a clear case of a small group of businesses having an interest in preventing interoperability. The companies make great profits, at high margins, from selling chargers that work for only a subset of devices. Other than a standard free-market argument that companies should be able to do what they like, there is no strong case for maintaining the existing mess of different cell phone chargers.

In a case like this one, a little bit of government intervention can go a long way to help consumers and, in this instance, the environment. The good news on this score comes from Europe. In 2009, the ten top cell phone manufacturers signed a memorandum of understanding (MoU) with the European Union to make cell phone chargers interoperable by introducing a shared micro-USB standard. The universal charger provides an easy way for consumers to share and borrow chargers. One obvious benefit of charger interoperability, and the largest driving force behind the standardization, is consumer convenience. Consumers save money, because they do not have to pay for several new chargers (one for home, one for the car, one for the road) each time they buy a new cell phone. And cell phone charger interoperability comes with clear environmental benefits. It reduces waste and increases the energy efficiency associated with producing the chargers.

The government's intervention was necessary, though, because certain market players made a lot of money selling the extra cell phone chargers required if the systems did not interoperate well. Companies often prefer to have proprietary products so that consumers must come to them for peripheral items, like power chargers. Cell phone service providers and handset makers both sell branded chargers at a markup. A common standard would allow generic providers to undercut these lucrative markets more easily than they can today.

Although the standardization of cell phone chargers was the result of a mutual agreement among the major producers of cell phones in Europe, government pressure also played an important role. The threat of top-down regulation from Brussels spurred industry players to cooperate and collaborate on developing a common standard and the MoU. This case shows us a perhaps unexpected way—legal scholars call it legislation by threat—in which governments can engage proactively to encourage interoperability.

Increased user choice is a key benefit of interoperability. By preserving or even increasing the diversity of options in the marketplace, interoperable systems, applications, platforms, and components allow us to test, mix and match, and mash up products and services to serve a specific need or purpose. Interoperability makes many of our experiences as consumers hassle free and more pleasant. At its best, interoperability can enable us, as users of information and communication tools, to do creative things. At an abstract level, we have the chance to become more autonomous as human beings.

Despite these benefits, getting to optimal levels of interop is often very hard when companies see keeping certain systems noninteroperable as in their own interest. Ordinarily, this viewpoint is right in the short term: if there is no generic alternative, companies can make extra money selling high-priced branded chargers every time they sell a smartphone, plus extra chargers at the airport to travelers who left them at home. But in the long term, companies may find that they benefit from growing entire markets by cooperating with others through interoperability strategies, delighting their customers in the process. Where companies do not provide interoperability out of self-enlightenment, the state may need to intervene to protect consumers.

C ompanies sometimes do, and sometimes do not, have an incentive to offer interoperable services and products. Consumers always have some degree of power in insisting on it. Customers can create pathways for interoperability despite the proprietary intentions of corporations. In the software, Internet, and digital media industries, the effect of consumer demand on interoperability has been substantial over the past decade.

Low levels of interoperability persist in many areas of our daily lives. Music recordings offer an instructive example. Growing digital sales have been eviscerating the market for sales of physical music media, particularly in brick-and-mortar stores. Tower Records shuttered its physical stores years ago, and Borders, too, failed as a big-box retailer of books and recorded music. Instead, a growing number of people download music, and much more, from online services such as Apple's iTunes or Microsoft's Zune Marketplace. Online distribution channels accounted for about 40 percent of total music sales in 2010.[2] Digital music sales reached $4.6 billion in 2010, and they continue to grow.[3] Over the past few years, online music services have exploded globally, with different underlying business models such as download- or subscription-based services.

These online music businesses offer most of their music wrapped in digital rights management (DRM) systems that regulate the various aspects of their services—often regulating them so that they cannot interoperate with other services. For instance, DRM systems determine the type and number of devices on which a user can play a downloaded or streamed song, the number of CDs on which a user might burn the song, or even whether a user can transfer the file to portable devices such as cell phones or MP3 players.

These DRM schemes could help make things more interoperable for consumers, but they generally work in the opposite way. DRM is used, in part, as a way to keep a consumer from buying a file on one system and playing it on another.

DRM was deployed initially to prevent piracy. The original rationale was that DRM fought copyright infringement. This approach blocked some consumers from pirating music, but it had little effect overall on the extent to which consumers turned to illegal downloading in general to obtain music. Over time, however, DRM schemes, such as Apple's FairPlay, have evolved into affirmative business strategies, often intended expressly to block interoperability. Merchants of digital goods now combine DRM systems with sophisticated terms of services, as Apple has, to create business ecosystems that do not interoperate completely with the systems offered by competitors.

From the start, the iTunes music store has been designed so as not to be broadly interoperable with other systems. The point of this purposeful non-interoperability has been to protect Apple's pole position in the digital media business. With only a few exceptions, Apple has refused to license its FairPlay DRM system to its competitors, including competing online music stores and device manufactures. Apple has basically decided to create its own, internally interoperable ecosystem rather than interoperating broadly with the services of other companies. As a result, its products (such as the iPod, iPhone, or iPad) and its music services primarily support content encoded with Apple's closed DRM system. Files in unprotected form, such as songs ripped from a CD, must be imported and "translated" into Apple's protected file format. Although the efficacy of DRM protection in preventing piracy remains unclear at best, it has allowed Apple to create and defend digital barriers that prevented its competitors from entering the successful iTunes/iPod/iPhone/iPad ecosystem.

Many consumers, and especially young users who expect to be able to move digital content around and share it with friends, dislike the restrictions that come with DRM schemes. Why shouldn't consumers be allowed to play songs purchased on iTunes on the device of their choice, including players marketed by Apple's competitors? Leaving aside the potentially negative impact on competition,[4] designers of noninteroperable online services and devices run a high risk in ignoring general consumer preferences and limiting consumer autonomy. In a world where high levels of interoperability are the norm, design decisions to make systems not interoperate with those of competitors may backfire. But where a system is integrated well enough vertically—in other words, interoperable with other attractive offerings from the same company, as in Apple's case—this interoperability strategy may work for a period of time.

In response to consumer demand, Apple has experimented with strategies that put a price on interoperability. Apple's iTunes Plus, for instance, offers many DRM-free songs for a higher charge per song than for DRM-protected ones. Unlike in the cell phone charger example, where consumers do not have the option of interoperability, consumers in the mainstream online music market currently can pay a premium and receive increased—but still not perfect—interoperability. This market-based way of ensuring interoperability is better than not giving consumers the option, but price discrimination may keep many people from enjoying the benefits of highly interoperable systems.

The digital music interop story appears to be headed toward a happy ending, at least from the consumer's perspective. The consumer demand for DRM-free music seems to be winning out. After six years of operation, Apple announced in 2009 that it would offer DRM-free music as part of its iTunes Plus program.[5] Amazon's MP3 store is among the leaders in the DRM-free music market. The store offers over 14 million songs that play on any device, including iPods, Zune players, SanDisk SANSAs, and players from other vendors, such as Sony and Samsung.

The interoperability story keeps repeating itself in the digital media field, in ways that stand to benefit consumers over time. The movie business has been trying to apply the painful lessons the music industry learned in experimenting with early DRM systems. The Digital Entertainment Content

Ecosystem, a consortium of over sixty companies from the movie and entertainment industries, consumer electronics manufacturers, and DRM vendors, has recently announced the introduction of a DRM platform called UltraViolet, which seeks to overcome all barriers that keep consumers from watching movies on more than one device. It is an impressively coordinated effort among otherwise fierce competitors.

The idea behind UltraViolet is to give customers a great deal of choice in how they enjoy the movies they buy, across a wide range of platforms. The UltraViolet service registers and stores users' purchases online and allows them to access the movies bought from an online or offline UltraViolet retailer on the device of their choice. For instance, a user who buys a DVD offered by UltraViolet can also stream the same movie on a smartphone or tablet.[6] UltraViolet-branded content will also be portable across game consoles, smartphones, Blu-ray players, set-top boxes, and mobile devices and can be shared among six registered account users.

The foundation of this new, promising DRM platform is a so-called digital rights locker, which stores information on a user's (or a household's) collection of purchased movies and allows them to be watched anywhere. Given the design and breadth of the service, UltraViolet may very well have a significant long-term impact on the market for movies and downloads. However, because Apple and its download store iTunes are not part of the consortium, UltraViolet will not run on important devices such as the iPhone and iPods.[7] The Walt Disney Company, which has its own competing technology called Keychest—which is also supported by Apple—has not joined the consortium and is not supporting UltraViolet either.

UltraViolet certainly has its limitations. Privacy will be a persistent concern, as it collects by default data that allows the service provider to know whether the user has visited a particular website, viewed content, or received a message.[8] And there are limits to what UltraViolet will empower consumers to do with the recorded entertainment they buy—for example, restricting editing or repurposing the content in creative ways—at least in its original design. (It will be interesting to see if an "UltraViolet Plus" emerges, as a follow-on to the iTunes story.) Despite these drawbacks, the development of UltraViolet demonstrates that consumer expectations can help drive cooperation in the marketplace to bring about relatively higher levels of interoperability.

I nteroperability offers consumers more than simply choice and autonomy. Interoperability can enable consumers to save time and to make better, more informed purchasing decisions. Many companies see their interests as aligned with those of their customers in this respect. If a company can make its services interoperable in a way that improves consumers' experiences, they may be likely to come more frequently to the store and spend more money while they are there. The innovative sales techniques of Starbucks Coffee Company offer an instructive example.

Starbucks is expert at reducing the amount of time customers have to wait for a drink. It has also figured out how to get customers to spend more than a dollar or two for a simple cup of coffee, even though they have to wait longer for the fancier, more expensive drink. Interoperability is part of the Starbucks strategy in this respect. In its US stores, Starbucks recently announced the nationwide rollout of the Starbucks Card Mobile payment program. Customers can now pay for their favorite Venti Chai Latte or Tall Cappuccino with their iPhones, iPod touches, or BlackBerries—or perhaps with reusable mugs that are preprogrammed with their beverage preferences and payment info.

The system works easily. Consumers first download the Starbucks Card Mobile app for their smartphones. Once they enter their Starbucks Card number, their phones essentially become their Starbucks Cards, which can be reloaded via PayPal or credit card. To pay with the phone, app users select "touch to pay" on the screen and hold the barcode that appears on the mobile phone up to the 2-D scanner at the register. Test runs with consumers suggest that the mobile payment program is the fastest way to pay for a morning latte at Starbucks.

The success of this program, powered by interoperability, is welcome news for consumers. The new system reduces the time they wait in line as they rush to get to work (or to bring coffee back for the boss). It is also good news for Starbucks, because the system makes transactions more efficient. Starbucks lovers use their smartphones while waiting in line, and baristas can focus on making the drinks rather than processing cash payments. Customers have little disincentive to buying the more expensive coffee that they really, really want but do not have (much) time to wait for.

As the Starbucks initiative shows, the combination of interoperability using barcode technology, web 2.0 applications—which allow interaction and collaboration among users—and mobile devices can make transactions more convenient for users and more efficient for retailers. But it also has the potential to empower consumers to make better purchasing decisions. For example, several applications for mobile phone operating systems, such as that of the Android, allow consumers to scan barcodes on products and look up prices and reviews. Some of these apps have built-in community features where users can find and follow friends, post products to their Facebook pages, share their own recommendations, and leave short comments. Applications like these are powerful educational tools in the hands of consumers. However, if business strategies using them are to function properly for the consumer, components and systems have to work together.

C onsumers want interoperability not just when they start using a new product or service but over time. Establishing interoperability is just the first stage. Maintaining interoperability is another challenge. Increasingly, we observe cases in which established interoperability unexpectedly breaks down, with a negative impact on consumers. Interop is not always forever.

A recent case from China demonstrates the importance to consumers of maintaining interoperability and illustrates how a company's quick shifts can break interoperability on the fly. Two heavyweight companies in China have been fighting one another fiercely, with interoperability as the battleground. One player is Tencent, the company that operates the hugely popular Chinese instant messaging (IM) tool QQ, which has a market share of about 75 percent and 636.6 million active user accounts. On the other side is Qihoo, a fast-growing provider of a variety of online security tools (including the popular 360 Safeguard software), which controls about 50 percent of the Chinese antivirus software market and has gone public on the New York Stock Exchange.

In September 2010, Qihoo's software detected that QQ (allegedly) engaged in suspicious spying activities by scanning irrelevant files on users' hard drives. It simultaneously announced a new piece of software, the 360 Privacy Guard, designed to detect data that QQ extracted from users' computers. Tencent responded swiftly by filing a complaint in a Beijing district court, alleging that Qihoo produced and disseminated false information and violated China's Anti–Unfair Competition Law. Tencent also stated publicly that it never attempted to access its users' private files. To retaliate, Tencent declared that its QQ client would "stop being compatible" with 360 products and would stop running on a system if QQ detected 360 software.[9] In response, Qihoo released the software add-on QQ Safeguard, which deactivated its rivals' detection function, removed advertisements on QQ's interface, and redirected QQ's links to its own website.

This highly publicized interoperability fight drew the attention—and intervention—of the Chinese Ministry of Industry and Information Technology (MIIT), one of the overseers of China's telecommunications industry. MIIT first criticized the business practices of both Tencent and Qihoo and ordered both companies to apologize to the public, to cease attacking each other, and to guarantee interoperability among their products.

MIIT then intervened a second time in January 2011. It unveiled draft Internet rules to regulate competition among providers of Internet information services. The draft Internet rules include provisions prohibiting providers from arbitrarily refusing to provide services, as well as provisions encouraging norms that would provide for comprehensive protection of online personal data. Most important, the rules include very broad provisions making it illegal for Internet service providers to take measures aimed at undermining interoperability with competitors' products. This rule holds regardless of whether or not the companies involved hold dominant market positions.

The fight between Tencent and Qihoo is a reminder of how easily service providers can adjust existing degrees of interoperability by remotely updating their software. Moreover, the Tencent-Qihoo fight demonstrates the possible consequences of this manipulation for consumers. Consumers faced the uneasy "choice" of either uninstalling software that protected their machines against viruses and spyware, or keeping that software and foregoing the use of the most popular IM client in China.

More interoperability is generally good for consumers, but it is not always good from the perspective of all companies. Companies and their customers are therefore not always aligned in wanting high levels of interoperability. Companies may use the ability to withdraw interoperability as a sword against their competitors in ways that cost their customers. A loss of interoperability usually limits users' autonomy and choice, even if it occasionally helps corporations' strategic purposes. Where breakdowns occur, intervention by governments can help consumers, as in the case of Tencent and Qihoo in the Chinese Internet wars.

I nteroperability among instant messaging systems in the United States is an instructive illustration of the way consumer demand has helped, with some success, to drive interoperability horizontally across platforms. The benefit to consumers has been improvements in the way people can communicate over digital networks. It took years, though, for consumer demand to lead to meaningful interoperability. The saviors, in this story, turned out to be neither enlightened corporate chiefs nor lawmakers but, rather, computer programmers from the open source community.

Instant messaging has become one of the most popular digital communication channels. IM is a real-time text-based tool used to facilitate communication between two or more people using IM clients. Examples of IM programs abound: Live Messenger, Skype, Yahoo! Messenger, Google Talk, Facebook's chat features, or Pidgin. In 2009, over 47 billion instant messages were exchanged daily from more than 2.5 billion IM accounts, owned by more than 1 billion users.[10] Instant messaging, also with texting, has replaced e-mail as a communication medium for many young people. Adults also use IM as an instrument to connect with colleagues in their professional lives or to keep in touch with friends. Today, interoperability among IM services empowers users to communicate in real time and across the globe regardless of the IM client their counterparts are using.

Use of IM increased sharply during the dot-com boom of the late 1990s. America Online (AOL) had an early and firm foothold in the IM market. Other big players included Microsoft and Yahoo! The growing competition among these three giants almost entirely displaced the smaller providers of IM services. In 1999, Microsoft unilaterally designed and introduced messaging software that was interoperable with AOL's. Not surprisingly, AOL, which controlled about 61 percent of the IM market in 2000, was less than pleased. AOL argued that Microsoft's new service led to "unwanted transmissions," and it took measures to prevent other IM providers from interoperating with its own service by actively preventing its rivals' servers from connecting to those of AOL. The controversy between the two giants escalated so far that the *Washington Post* referred to the conflict as "electronic warfare."

The dynamics changed dramatically with the merger of AOL and Time Warner in 2001. When the Federal Communications Commission (FCC) reviewed the AOL–Time Warner merger, it mandated that AOL stop blocking the attempts of its competitors to interoperate with AOL's IM software. Furthermore, the FCC requested that AOL take the initiative to pursue interoperability with other providers. Not only was this mandate significant for consumers, but it also marked the first time the FCC regulated a software application born out of the Internet. The FCC argued that there were network effects at work in the IM market. As a consequence, the FCC contended, it needed to mandate greater levels of interoperability in order to increase competition and to empower consumers. The AOL IM case is another instance in which the role of government was key in establishing a more interoperable ecosystem.

Within only a few months of the FCC mandate, AOL's market share dropped, while Microsoft's and Yahoo!'s increased. At the same time, a new software application by the name of Trillian entered the market and quickly became popular. Trillian functioned as a kind of "meta" service. It let consumers connect to multiple instant messaging services at once, using, in effect, simulated interoperability protocols. Originally these services included AOL's Instant Messenger, ICQ, and MSN Messenger. Trillian could not make the systems all talk to one another, but it could enable a consumer to see them all at once. Using Trillian meant that the user only needed to open one application for IM, instead of several different clients. Trillian aggregated all the contacts from the IM services with which a user was registered into one single list that the user could group in various ways. This kind of interoperability is called as-if interoperability, because it was built on top of systems that were still noninteroperable among each other.

In light of the emerging trends towards higher levels of IM interoperability, the FCC lifted its prior mandate. Ironically, only a few months later, Microsoft and Yahoo! announced software upgrades that blocked third-party IM clients, including Trillian, from communicating with their users. The companies cited security and privacy concerns in justifying their respective blocking efforts. These were the same arguments that AOL had unsuccessfully used to support its initial actions designed to restrict interoperability.

The IM war among the market leaders continued for years. Against this backdrop, consumer demand for IM interoperability exploded. Big players in the financial industry, which started to rely heavily on instant messaging, got organized and called for interoperable corporate IM software. In response to the increasingly complex uses of IM by the financial sector, new players, including Google, entered the IM market. Incumbents in turn established alliances in the hope of maintaining a proprietary edge. The net result was intensified competition, but also increased interoperability across services. In fact, several of the core functions that corporate users demanded required increased interoperability among the competing platforms.

Consumer demand can also breed successful workarounds. In the case of IM, computer programmers turn out to be big IM users. One way to work around corporate software makers fighting to keep systems proprietary is to start over from scratch using open source tools. In the case of IM, the open source world produced Jabber, an open technology for instant messaging. The basic protocols of the technology had already been developed and released in late 1990s. The growth of these open technologies has led to an open IM standard known as the Extensible Messaging and Presence Protocol (XMPP). The Internet Engineering Task Force (IETF) and others in the open software movement played an important role in this standardization process.[11]

In response to the growing consumer demand for interoperable services, an increasing number of IM platforms now support XMPP. The proliferation of this standard laid the groundwork for consumer-empowering interoperability across various private and public networks, including popular services such as AOL's Instant Messenger, Apple's iChat AV, Microsoft's Live Messenger, Yahoo!'s Messenger, and many others. The net effect of this open standard is much greater interoperation among IM services. In the case of IM, the combination of consumer demand and open source innovation led to the change that the corporations and government were taking quite a while to get done.

The most fundamental shift made possible by interoperable technologies is the way they enable us to be not just passive consumers of information but also active *creators* and reusers of content in a public, networked environment. Digital technologies, rendered interoperable across platforms, empower consumers in unprecedented ways to express themselves through content creation and content sharing.

Content creation on the Internet is particularly popular among younger users, though adult users have caught up recently. According to a 2010 Pew Internet report, 38 percent of Internet-using teens in the United States shared their self-created content online.[12] User-created content (UCC) includes everything from status updates on social networking sites to blog posts, fan fiction, photos, podcasts, self-made video clips, comments on websites, virtual-world items, and so on. In many cases, user-created content is relatively trivial (such as a tweet like "at the coffee shop" or "landed in SIN"). In other instances, such as political mashups, creators are aiming for much broader impact.

Online content creation by consumers takes place largely in and across social networks. Social networking sites have generally achieved high levels of what we call *vertical* interoperability but much less interoperability across systems, or *horizontally*. For instance, one may access Facebook on all popular browsers and via most types of smartphone operating systems available in the United States and Europe. But social networking sites have little incentive to foster real horizontal interoperability that would allow users to interact across different platforms.

Although it is relatively easy for any single user to cross-post from a service like Twitter or WordPress (a popular blogging platform) to Facebook, it is currently impossible for a user's friend to leave a response that will then also appear across platforms. Furthermore, interoperability that flows one way can also lead to lock-in effects. For example, Facebook makes it easy for users to develop their own communities by helping them draw on the data provided by their Gmail accounts, their friends' profiles, and other information. However, it makes it almost impossible for them to easily take that accumulated community with them when they go—say, to try out Google+, a competing service.

Online creators also want to take the content they have created on one platform and transfer it to another, but they often end up frustrated because of the shortage of horizontal interoperability. Consider massively multiplayer online role-playing games (MMORPGs) such as Ultima Online, EverQuest, or World of Warcraft and social spaces like Active Worlds, There, and Second Life. Many of these virtual worlds are not only extensively customized but to a significant extent even created by their users, who are represented by avatars that move and interact in the shared spaces. Although virtual worlds such as Second Life may have lost much of the luster they enjoyed in the past decade, online virtual worlds are being created at a faster rate than ever before. In 2007, fewer than 150 such platforms existed. By 2011, over 450 virtual worlds had cropped up. And they are reasonably big business: together, they generate roughly $4 billion in revenues per year, according to a report by a specialized consulting firm.[13]

Interoperability was not built into virtual worlds in their original designs. They were more or less closed universes because there were few standards defining basic functions in virtual worlds, such as how to move an avatar in a shared space and how to manage its interactions. Without such standards, it was almost impossible for users to "travel" from one universe to another or to exchange self-created virtual-world items across platforms. This lack of interoperability limited the autonomy of users and put limits on the growth of virtual-world platforms overall and in the long run.

Consumer demand has led companies in the virtual-world industry to come together to discuss the possibilities for horizontal interoperability. The industry has started to collaborate on and experiment with different interoperability designs, leveraging existing standards for protocols and content formats. The goal is to empower users and allow them to move their avatars from one world to another and transfer user-created virtual-world items across services—and in the process, to grow the ecosystem of virtual worlds through interoperability. The net result of this type of collaboration is almost certain to be good for the virtual-worlds industry, as well as for their communities of consumer-creators.

Higher levels of interoperability, especially *horizontal* interoperability, are generally good for consumers. An absence of interoperability ordinarily leads to negative consequences for consumers. Interoperability can increase the ease of use of information and communication systems, and it can let users mix and match and mash up products and services and customize them for their own needs. Interoperability does not just make our lives, as consumers, more convenient and flexible. In some cases, interop enables us to make better decisions by increasing transparency and access to information about products.

Most important from a societal perspective, interoperability empowers us in our role as active users of digital technology. The powerful role that interoperability plays in this respect becomes clear when we look at the ways in which millions of computer and smartphone users now can create and share digital content. Whether we take or share pictures, create videos, post status updates on our preferred social networking sites, or comment on friends' blogs, interoperability is the key enabler of creative use of social media.

Consumer empowerment is among the important reasons why customers, companies, and governments should work together to achieve higher levels of interoperability—in general, at least. But there are major roadblocks. Companies often think their interests lie in keeping their products proprietary rather than making them interoperable. The kind of interop that matters to consumers can take a long time to emerge because of these competing corporate interests; it is sometimes hard to achieve, and it can be undermined or lost overnight. Sometimes companies or computer programmers will come together, especially when prompted by their customers, to seek higher levels of interoperability through collaborative efforts, as in the case of Instant Messaging or the virtual-worlds community. But sometimes market forces lead to fights in which companies reduce interoperability levels. In these cases, accomplishing optimal levels of interop, especially horizontally across platforms, often needs a nudge—or more—from the government. Customers, and the majority of market players, stand to benefit from the interoperability that ensues.

CHAPTER FOUR

Privacy and Security

W e do not want complete interoperability at all times and in all places. In certain contexts, it can introduce new vulnerabilities, and it can exacerbate existing problems. Privacy and security risks are, of course, the primary concerns when it comes to interoperability. An interoperable system has, by its nature, more points of open access to data. These points of access render systems vulnerable to bad actors—ranging from marketers to hackers with ill intentions—who can potentially access and misuse these data, inject malicious code, or otherwise compromise systems.

Sony learned this interoperability lesson the hard way. Sony discovered in April 2011 that 77 million PlayStation Network (PSN) service user accounts had been hacked. The problem, it turned out, was that its networked gaming system relied on technical and user controls that were not well established. The hacker had obtained user addresses (city, state, zip code), e-mail addresses, birthdates, PlayStation Network passwords and usernames, online IDs, and profile data, as well as possible access to purchase histories, billing addresses, and even credit card numbers. If the PlayStation systems had been stand-alone units, unconnected through a network, the security levels on the systems would not have mattered. Of course, interoperability was not the problem on its own—but it became a problem because the security and privacy controls were inadequate.

The vulnerability of privacy and security within networks is a consequence of the increased *complexity* of an interoperable framework. It is not interoperability per se that gives rise to increased privacy risks but, rather, the specificities of its implementation. The level of privacy and security protections within an interoperable system depends on how we design, implement, and monitor systems at the technology, data, human, and institutional layers. Unfortunately, the information and communications field is full of examples in which companies, in particular, have failed to meet this fundamental design challenge.

The way to frame this design challenge is to focus on the degree to which systems are made to interconnect—the *degree* of interoperability. Interop is not binary. There is no single form or optimal amount of interoperability that will suit every circumstance. Just as interoperability operates at multiple layers, there are many degrees—in fact, infinite degrees—of interoperability. The problems of privacy and security help bring the importance of this dimension into relief.

Most of the time we want an optimal, not a maximum, degree of interop. It is not always desirable to have complete interoperability in every circumstance when pursuing goals such as consumer empowerment or innovation if, at the same time, we want to protect core social values like privacy and security. Lower levels of interoperability—think of it as incomplete interoperability—are often desirable. Sometimes friction in a system is in the public interest. Sometimes it is best to have no interoperability at all— a perfect *lack* of interoperability. And sometimes, where interoperability exists, we have to be prepared to address, through markets or through regulation, heightened privacy and security risks.

S ecurity systems offer instructive examples of how limitations on interoperability can be desirable. Security systems are designed to keep people out of certain places at certain times. A city dweller might leave his door open during the day to invite neighbors to wander through the house during a block party. The same city dweller might triple-lock that same door that same evening while he is asleep. The complex relationship between security and interoperability helps illuminate some of the drawbacks and limitations to complete, omnipresent interop.

Interoperability does not automatically make systems less secure. The fact that a given computer system can interoperate with others does not mean that more people have access to underlying data in the system, for instance. However, depending on the implementation at the different layers and without sound security measures, increased interop among systems can add to the vulnerability of the different components or systems. The opposite also might be the case: systems with higher degrees of interoperability can be rendered more secure than systems with low degrees of interoperability. There can also be high security risks associated with systems that are not at all interoperable.

Security problems are related not to interoperability itself but, rather, to what interoperability makes possible. This distinction makes a difference, especially when it comes to designing interoperable systems. Fundamentally, any system that has more points of open access—for example, one that makes it easier for people to come and go—might give rise to security problems. Put another way, in a highly interoperable system, intruders might be able to enter a space unencumbered and take something to which they are not entitled.

In the information-era version of the same example, interoperability might mean that data can flow through and across systems without restraint. As such, interop also might make it possible for a bit of malicious computer code—known as badware or malware—to slip into a computing system and introduce a virus. Or, to give another example, interop might make it possible for information a teenager has uploaded to Facebook to move from that context into another that the teenager did not mean it to reach. The problem of sexting works this way: a teenager sends a nude picture of herself to a boyfriend, who in turn sends the picture to his friends, who move it from their mobile phones to the web and beyond after the nasty, inevitable breakup. In all these examples, we might be tempted to get the law involved.

This complex relationship between security and interop reveals the fact that the optimal degree of interoperability varies by circumstance. Walls between systems, or rules that function like walls in certain cases, can serve important purposes, especially from a security perspective. Sometimes we want these walls to be permeable, other times not. These policy decisions should drive the way we design interoperable systems. In a digital environment, there are often sliding scales, so that reaching an optimal point of security is at least theoretically possible in some circumstances.

Every once in a while, it makes sense to engineer systems to allow no interoperability at all. Very often, the reason for designing systems in this noninteroperable way is to guard against human error. The example given in the Introduction of the different size nozzles at the gas station—for diesel or regular fuel, neither of which can work with the fuel tank aperture of the wrong type of vehicle—makes this point. This example points not to a drawback of interoperability but to an instance where it is better not to have interoperability at all. There is no foreseeable instance where the driver of a car that takes ordinary fuel would want to put diesel in the fuel tank. To do so would be to risk ruining the engine. At the same time, there are millions of transactions per day in which a consumer, without thinking about it, might reach for the wrong pump at the gas station and run an unnecessary risk. It is smart to keep certain systems from interoperating in order to prevent harm.

More often, situations call for limiting the degree of interoperability in order to accomplish certain goals, such as security. Systems are often designed to work best with a certain degree of friction built in. In such cases, it might be a good thing to make the systems work together in a fundamental sense, at the technology and data layers, and then introduce policy or legal requirements that the parties take certain steps before the systems are permitted to interoperate.

By using sound authentication mechanisms, we can design highly interoperable systems that are also secure. Someone seeking access to a certain physical environment might need to show a series of credentials before being admitted, for instance. Consider the case of a high-security research facility. Visitors are required, by rule, to call ahead to have their names put on the list of people expected on a given day. When they arrive, they are required to show certain kinds of identification. They might have to submit to retina scans, for instance, prior to being granted full access to the facility.

The design of this high-security research facility builds in friction, which is meant to allow interoperability in some cases but not in others. This design principle is *selective interoperability*. Operability is not meant to work for everyone all the time, by design. The idea is that one can permit certain things—such as potentially dangerous or especially sensitive research—to occur in a certain place and enable some people to have complete access, some people to have limited access, and some people to have no access at all.

There are important variants to this concept of selective interoperability. Contrast this approach of selective interoperability to the web services context, where we often see unilateral openness. For instance, a web services provider like Facebook or Google voluntarily creates an open application programming interface (API) that allows anyone to interoperate with its services without the need for further approval or cooperation. Google Maps or its Android platform are examples of systems where developers do not need to ask permission before they start to build. Engineers unrelated to Google are allowed to write code that interoperates so long as they agree to certain legal terms, often a simple click-through agreement with no lawyers involved. The only friction involved is the decision point on the part of the potential participants: the developer must decide whether or not to accept Google's invitation to interoperate, that is, whether the terms are acceptable or not.

It is important to distinguish selective interoperability *by design* from instances in which the exchange of information is limited *in fact*. In the context of office documents, for instance, a computer program puts up a warning screen that says data will be lost when a file is converted from one file format to another. The user may or may not care about what is lost in this transfer. The point is that the systems are not in fact fully interoperable:

some of the data, or data about the data (called metadata), does not flow from the first version to the second version.

This scenario is distinct from the access situation just described, in which all the data can flow from one system to another. In that case, though, *by design*, the system's architects decided to build organizational or procedural barriers making it harder to transfer some or all data across the systems. The high level of interoperability at the technology and data layers in the strict sense is preserved. The limitations occur at another level: in select cases, complete interoperability is not available to the person seeking access. The law or policy decisions, such as lists of approved people who may enter, can function as limits on an otherwise highly interoperable system.

This distinction is important for two reasons. First, it highlights in a precise way what interoperability in the information context means: it is about the flow of data across systems. Second, it exposes design options for addressing drawbacks to interoperability as applied. These design principles do not necessarily mean we have to accept a lower level of interoperability across the board. The example of calling before arriving at a high-security facility makes this point: rather than building permanent barriers, one might, for example, want to be able to give the chief scientist of the institution full access to the facility on the day she visits from the home office.

A variant of selective interoperability is *limited interoperability*. A system might be established to provide only partial interoperability to other systems or interoperability only in certain contexts. Think of the chief scientist coming to the research facility. The access system is designed to allow that visitor access to part but not all of the facility. There are rooms or labs to which she will not be admitted. This is limited interoperability.

Sometimes, the ideal system works like a screen door: it lets in the light but keeps out the flies. There are options as to what functions as the filter. The technology might allow both light and flies to pass through the aperture. Instead of using a screen door as a filter, we could require that a person with a flyswatter be stationed beside the doorway. His job would be to swat all the flies as they go through. The example suggests the upsides and downsides of both options: the screen door might be more consistently effective at keeping out flies, but it might limit the flow of more than just the flies; the man with the flyswatter might only hit flies, but he would surely miss some, which would enter the house.

The many ways to accomplish interoperability come with strengths and weaknesses. The most highly interoperable systems are very often the most permissive of information flowing among them, but that need not be true. Some computing systems are highly interoperable but are built with very high levels of security. Bank ATMs exemplify a system that is highly interoperable in terms of letting customers from virtually any financial institution take out cash from any other financial institution. And yet the system overall is highly secure. The choice is not whether or not to make systems interoperable but, rather, how to manage the flow of information within interoperable systems to ensure the desired level of security.

T he loss of privacy is, rightly, what many of us fear most when it comes to interoperability. But it is important to remember that interop does not automatically, by its very nature, operate at cross-purposes with individual data privacy. A perfectly interoperable system can be used in such a way that no one's privacy is violated at all. One might even be able to imagine interoperable systems with privacy-enhancing qualities. But in practice, that is not usually the case. High degrees of interoperability among consumer-facing systems make it easier for someone to violate someone else's privacy. High degrees of interop can also make it more likely that information about a person will be shared with others.

Time after time, companies have had to face the fact that consumers get very unhappy when certain kinds of interop are introduced into systems without full disclosure or explanation. Google and Facebook both had highly publicized problems of this nature within about a year of one another. This privacy-threatening interop can occur within a given firm (Google) or among firms (Facebook). In both instances, the company involved decided to create interoperability of a sort that got it in trouble. These interoperable systems resulted in the exposure of individuals' data in unexpected contexts. The customers involved felt that the companies had violated their privacy: they said they had never agreed to that kind of information sharing.

Google's Buzz, introduced in February 2010, is an example of problematic interoperability within a single firm's products and service. Google's Gmail, which offers free, web-based e-mail with a large amount of storage capacity, has proved enormously popular. For many users, a Gmail account becomes the core of a series of related services they use at Google. Gmail allows users to create an account with Google that can be connected, seamlessly, with other Google-owned services. All the cloud-based office productivity services, such as Google docs, can be connected to this single account. A user's YouTube account, too, can be connected to this Gmail account. As part of this process of integration, many users also integrate their contacts into Gmail, either through an explicit import process from another program or through an implicit process of corresponding with others. Through whichever process, users of Gmail can develop extensive lists of contacts. So far so good: by most accounts, all this interoperability made Google customers' lives easier.[1]

Then along came Twitter, the microblogging platform that took the world by storm between 2007 and 2010. Facebook, too, became increasingly popular during this time, in fact surpassing Google as the most commonly used technology platform in the United States.[2] Both Twitter and Facebook enabled users, through the web or via a mobile device, to post short statements about their whereabouts or to provide comments online. These short messages would be accessible either to their "friends," in Facebook's terminology, or to those who decided to "follow" them, in Twitter's parlance. Google thought users would want such a service connected through Gmail. Google engineers designed a new product, Buzz, to meet this demand.

The problem with Buzz was not the product itself, nor the fact that it was made to interoperate with other parts of the Google suite of online services. In previous cases, customers had seemed to appreciate, for example, the ease of accessing a spreadsheet or a text document in the Google cloud from Gmail. The previous efforts by Google's engineers to render their systems internally interoperable had not been especially problematic from a user perspective.

But a furor broke out shortly after Buzz was introduced. The story is complicated and a bit murky. Users complained that Buzz automatically generated a list of "followers" for each Gmail user, based on the user's most frequent e-mail contacts. According to the court filings, these contacts were displayed in "follower" and "following" lists that were visible to Gmail users' automatically generated Buzz followers and, in some cases, to the public. So, for instance, before Buzz was introduced, there would be no easy way for other people to see the contacts of a given Gmail user (call her Alice). Let's imagine that Alice has a contact named Bob. After Buzz was rolled out, plaintiffs claimed, the names of Alice's contacts, including Bob, might be visible to a third Google customer (call her Claire), without either Alice or Bob having adequately consented to the disclosure.

Further complicating matters, users complained that these lists were generated without notice to, and without the explicit consent of, either user. Google's engineers swear that the services revealed nothing about a user who did not opt in to, or choose to turn on, Buzz. According to Google, Alice's contacts would never be revealed without Alice's explicit agreement. Furthermore, it argued that even for those who did choose to use the free service, nothing was done wrong after a user began to use the service. Google claims that the way the program works was spelled out in writing to the user in sufficient detail that Alice should have known exactly what was going to happen.[3]

Many users did not see it Google's way. A group of Gmail users, with Harvard Law School professor William Rubenstein as one of their lead lawyers, filed a class-action lawsuit against Google. The plaintiffs alleged that Google violated a series of US privacy-related laws through the way they introduced Buzz. At more or less the same time, privacy advocates at the Electronic Privacy Information Center (EPIC) filed a complaint with the Federal Trade Commission (FTC) against Google on similar grounds.[4]

The focus of user complaints was not the functionality of Buzz itself but the way it interoperated with other Google services—in this case, the contacts feature of Gmail—and the exposure of data (such as Alice's relationship to Bob) outside the original context in which Alice put Bob into her list of contacts. To be clear: the problem was not the interoperability as such. The Buzz system might be made to work perfectly from an interop perspective with Gmail and its contacts without giving rise to alleged privacy violations. The problems stemmed from the rules that made this interoperability kick in without, in the plaintiffs' view, sufficient notice about what it would allow others to see in users' contacts lists.[5]

Privacy concerns have emerged as Google's Achilles' heel, with interoperability playing a big part in that development. The short, painful story of the Buzz rollout makes this exposure obvious. Google quickly made changes to its Buzz service after the uproar, and within months the parties moved to settle in federal court. Under the terms of the settlement, Google agreed to pay $8.5 million but admitted to no wrongdoing. Google also made significant changes to its privacy policies to settle related charges filed against it by EPIC with the FTC. Ultimately, the Buzz rollout's total costs to Google will be much higher when one factors in legal fees, the cost of public relations firms, and the losses from reputational damage and continued erosion of trust among a class of users.

Facebook experienced a similarly unhappy experience with the rollout of its Beacon service about two years before Buzz. Beacon is an example of interoperability gone awry, not within a single firm's services but across multiple firms. Facebook enables users to express preferences about people, institutions, and products through a variety of mechanisms on its site and, increasingly, across the web. For instance, a Facebook user can indicate that she "likes" a status update that one of her friends posts online. When President Barack Obama posts to Facebook a short statement about the importance of establishing a clean energy future or of diminishing US reliance on foreign energy sources, thousands of users (or more) might click on a "like" button to indicate their agreement with the president on this topic.

Facebook has emerged as the most popular web service in its core market, the United States. It is quickly growing in popularity in markets around the world. Unlike Google, Microsoft, or Yahoo!, though, Facebook does not have an obvious way to make money from the service it provides. Many Facebook users sign up for the service and never click on the advertisements rendered along the right-hand side of many pages on the site. For those users, Facebook was functionally free; the company captured little value directly from these nonpaying, nonclicking customers. In the long run, Facebook is amassing value from these new sign-ups, of course. As cofounder and CEO Mark Zuckerberg correctly recognized, the huge value in the service rests in the "social graph" that connects hundreds of millions of Internet users to one another. That said, Facebook still needs to find ways to make money from its popularity. The rendering of contextual ads on most pages might make the company profitable, but not as profitable as some of its competitors.

In November 2007 Facebook's executives were determined to be more creative in extracting profit from their product. One of Facebook's ideas for a new revenue stream was to make its system more interoperable with companies that marketed products and services to Facebook users on the web. Beacon, as the service became known, was a program whereby Facebook teamed up with Internet retail companies to make their systems interoperate in order to generate higher profits for both firms. In total, forty-four partner websites, including Overstock.com, eBay, Fandango, Blockbuster, and many others, participated in the Beacon program.

The system was simple in its design, but it sounded pernicious to users when they came to understand what it did. When a Facebook user made a purchase on a partner website, such as Overstock.com, a marketer of discounted merchandise, a few bits of data would be able to flow back to Facebook. Facebook would then render a short notice on the user's Facebook page showing the product he or she had just purchased. This notification could also appear in the news feeds of the user's friends, depending on how the user had configured his or her settings on Facebook. The interoperability at work here: the flow of data about the user from Facebook to Overstock.com, and about the product purchased from Overstock.com back to Facebook.

Just as in the Buzz case, Beacon caused an uproar among web users and privacy advocates. The problem again was not the interoperability among systems but, rather, the exposure of data from one context in another context. The system was also based on a premise called "opt in by default," which meant that Facebook automatically enrolled users in the service unless they explicitly said they did not want to be part of it (which sounds a lot like "opt out" rather than "opt in"). Users expressed myriad concerns, ranging from the fundamental (alarm at the continued erosion of user privacy on the Internet) to the worrying but probably not common (fear that a racy item purchased for a lover by an adulterous husband might appear on a news feed and be seen by his spouse).

A class-action lawsuit ensued in the Beacon case, too, which resulted in a settlement of $9.5 million. For a small, fast-growing company like Facebook, the reputational hit and the hassles associated with the lawsuit represent a far greater loss than the monetary cost of the settlement. Privacy concerns about interconnected web services will only continue to grow and will threaten companies like Facebook and Google if they do not figure out how to address concerns before they blow up into class-action lawsuits.[6]

The problem in the Beacon and Buzz cases stemmed from the way Facebook and Google made their respective systems interoperate and from their failure (their adversaries claim) to notify customers about how data would be used. It is not a foregone conclusion that we will have less privacy and security in a highly interoperable world. But we need to be vigilant about how interop is introduced, in what degrees, and by whom, in order to ensure that we do not give up more than we gain as we progress.

Higher levels of interoperability are not always good, in large measure because they can give rise to problems of security and privacy. One way to avoid these problems is through design. Facebook and Google might simply refrain from designing products and services that interoperate in ways that lead consumers to feel that their privacy has been violated. Sony has no doubt already tightened the security controls on its networked PlayStations since its interop-related debacle.

The largest problem with interop and privacy has to do with businesses that link data about consumers without asking anyone at all, where an especially intrusive form of interop is core to their business model. Massive data aggregation firms such as ChoicePoint should stop drawing data from so many sources in order to enable others to market goods and services to us more effectively. At least in the ChoicePoint example, and to some extent in Google and Facebook examples as well, though, the companies involved generally do not want to design their systems to protect consumers. They have other drivers in mind: low costs, high-margin sales, protection of their competitive position, and so forth. ChoicePoint, and many others, exists *in order* to collect (and then sell) data about consumers; it is hardly realistic to expect the company to abate its core mission for the sake of protecting privacy.

The market will not take care of privacy and security issues on its own, and therefore the law plays an important role in managing this aspect of interoperability. First, and most fundamentally, background rules establish a playing field for companies that work with personal information. There are limitations, established in law, as to what companies can do with an individual's personal information. These limitations tend to be stricter in the European Union than they are in the United States, but in both cases there are legal rules. States occasionally set up specific laws governing particular forms of information: for instance, doctors cannot share information about their patients in an unregulated fashion, which necessarily limits interoperability but produces other benefits. And the law can create a "right of action" for interventions by private parties, as in the Google and Buzz cases, or by state agencies such as the FTC, when the public interest is at stake.

A good example of the state stepping in is the role the FTC played in the Buzz case. At the same time that the private lawsuit against Google was underway, the FTC investigated Google's practices. (EPIC's formal complaint against Google had prompted the agency to act.) The FTC's investigation of the matter was followed by a broad settlement of privacy-related concerns in March 2011. The consent order required Google to "establish and implement, and thereafter maintain, a comprehensive privacy program" to protect consumers.[7]

Security and privacy present clear cases for the need for specific rules that set limits on interoperability. In other examples, we have seen the need for state intervention in order to prompt interoperability in the first place.

These multiple examples come together in a principle about the law in general. The law can function in three productive ways: it can serve as an enabler (of beneficial forms of interoperability, as in the case of public safety and emergency communications), it can create a level playing field (to create a competitive market through interoperability, or so that interoperability can thrive), and it can function as a constraint (to ensure that interoperability does not lead to unwanted effects, such as privacy and security problems). If deployed with skill, the law can play a central role in ensuring that we get as close as possible to optimal levels of interoperability in complex systems. Against the backdrop of reasonable rules, companies can and should design interoperable systems that do not create these problems in the first place.

CHAPTER FIVE

Competition and Uniformity

C ompetition is a good thing in markets for information technologies because it often leads to innovation and greater diversity of services to customers. These positive outcomes occur when there is a level playing field and low barriers to market entry. A major argument in favor of interoperability is that it tends to promote competition by making it easier for new players to participate, especially in information-related businesses. Over nearly a decade of studying interop, we have analyzed many examples in which interoperability has played an important role in leading to competition in the marketplace. For instance, in social media, a broadly interoperable ecosystem of technologies has led to an explosion of applications built on platforms like Twitter and Facebook. That same interoperability has helped bring about a similar explosion in applications for mobile devices, building upon the operating systems for Apple's iPhone and smartphones that run on Google's Android platform.

By and large, whenever the level of interoperability among technologies in a given market has increased, competition and innovation have increased as well. Our observations are backed up by standard economic analyses. Economists who have studied this issue often argue that increased interoperability is likely to foster competition and innovation by reducing lock-in effects (in many, though not all, cases) and by lowering market entry barriers.

That said, although more interoperability is *generally* good for competition and innovation, the story gets much more complicated as we delve into it in greater detail. Interoperability sometimes leads not to more competition but, instead, to anticompetitive situations. Imagine two very large information technology companies that decide to work together to make their software interoperable with one another's systems but not with those of other competitors. Over time, these two very large companies together gain a large share of the market. This growth in market share for these two companies has come about through their interoperability agreement, which has made it easier for their customers to run their combined systems. The net effect, over time, might be to crowd out other technology companies in the same line of business serving similar corporate customers. In this example, the degree of interoperability in the marketplace has gone up, but the degree of competition has gone down.

There is a second wrinkle to the general rule that interop breeds competition. Even in the general case where more interoperability leads to more competition in the market, the net effect is not always maximum innovation. According to one strand of economic theory, firms may have an even stronger incentive to be innovative in circumstances where low levels of interoperability promise high profits to any company that beats all its other competitors soundly. The promise of monopoly returns, where systems are not at all interoperable, may provide the highest incentive of all. A company may perceive that the best way to get the highest return from its investment of time and capital is to innovate in such a way as to develop a noninteroperable system that becomes enormously popular. The early iTunes Music Service is an example of such a strategy. Apple's innovations plainly broke open the market for digital music, but they did so in a way that involved very little interoperability (at least initially) with the services of their competitors.

Finally, it is possible for interop to lead to the use of a single technology over a long period of time, which itself may have negative effects. The two potential problems here are uniformity and lock-in. Uniformity may or may not be a bad thing: it is possible that it makes sense for everyone to be using just the same technology to get a certain job done, and the innovation happens at higher layers of the technology stack. But it also can be undesirable. Diversity among systems that work together but are not necessarily the same can ensure that innovation continues along multiple fronts. Diversity within systems can help prevent lock-in over time, as the relative merits of various technologies can easily be compared side by side. This is where the distinction between complete standardization (in which the same technologies are adopted across the board) and interoperability (in which technologies merely need to work together in certain ways) is so important.

The policy challenge is to set the balance right so as to encourage the kinds of interop that drive competition. At the same time, the policy environment should disfavor those forms of interop that can lead to uniformity or to lock-in. This requires, first and foremost, a deeper theoretical understanding of the complicated interplay between interop and competition, of how the behaviors of individual companies shape this delicate balance, and of what governments can and should do about it.

A lack of interoperability among products and services marketed by powerful tech companies is, largely, bad for competition and innovation. As economists argue, higher levels of interoperability tend to increase competition and foster innovation. Many case studies in the information and technology business back up this argument. Government regulators have generally agreed and have intervened where necessary to promote competition. The essential problem is how to accomplish the right level of interop and how to avoid ending up with uniformity instead of competition.

The long-running dispute between Sun Microsystems and Microsoft and the intervention of the European Commission in the matter help illuminate the challenges associated with getting the interoperability-competition balance right. In the fall of 1998, software giant Microsoft received a letter from Sun Microsystems, an archrival. Sun was requesting technical information from Microsoft regarding the Windows operating system. Sun wanted to know details about how Microsoft's core product, the primary driver of its vast profits, worked under the hood. Sun wanted to develop an operating system for servers that could work with Windows, and its engineers needed the information, they said, to help them make their system interoperate with Microsoft's.

Microsoft's response to Sun was unequivocal: no. Microsoft's executives at the time presumably decided not to share the information that Sun wanted because it would be better for Microsoft not to let Sun know about the inner workings of its cash cow, the operating system. Microsoft must also have thought that there was not much benefit to letting Sun develop technologies that would interoperate with the Microsoft operating system in the way that Sun's executives had in mind. And Microsoft may well have bet that the regulators in Europe would not go so far as to mandate disclosure of technical information from one rival company to another. After all, the law in the area of computers and interoperability is still not well developed, and it was much less so in the 1990s.

Microsoft's refusal to share technical information about Windows triggered the fiercest of interoperability battles, which would last over ten years. Regulators in the European Union took special interest in Microsoft's refusal to turn over this interop-related information to Sun. Saying that they were protecting European consumers, the European Commission brought a legal action against Microsoft. This action resulted in a record-breaking fine of €497 million imposed on Microsoft by the European Commission and a 2007 landmark decision by the European Court of First Instance on the matter. (The parties ultimately resolved the matter once and for all in a settlement in 2009.)

In a complex and controversial decision, Europe's second-highest court ruled that Microsoft had violated EU competition law by abusing its dominant market position in PC operating systems in two distinct ways. First, the court ruled that Microsoft had illegally refused to provide computer protocols that would enable a certain category of competing operating systems (for work group servers) to "talk with" Microsoft's Windows operating systems. This part of the decision was related to Sun's original request for interoperability information. Second, in a related matter, the court agreed with the European Commission that Microsoft violated EU law by bundling its Windows Media Player with the Windows operating system. Because Windows was on more than 90 percent of all PCs at that time, bundling the products together stifled the competitive ability of rival media players.

It was not the first time Microsoft had run into problems with antitrust authorities on the subject of interoperability policy. A few years earlier, in May 1998, the US Department of Justice (DOJ) and the attorneys general of twenty states had filed a civil antitrust case against Microsoft. The government entities argued that Microsoft had abused its monopoly power by bundling the Internet Explorer web browser to its Windows operating system. The case was settled in 2002 with a landmark agreement between the DOJ and Microsoft that fundamentally altered the way Microsoft competed in the marketplace. The net effect of these struggles between Microsoft and the competition authorities in Europe and the United States has been higher levels of interoperability among a wide range of information technology services and products. Microsoft, too, has changed its strategy dramatically since that time, becoming one of the biggest boosters of pro-interoperability approaches to technology development.

To understand the relationship between interoperability and competition, it helps to take a few steps back in the story of Microsoft's history. Beginning in the 1970s, Microsoft developed a suite of software products that eventually gave the company control of over 90 percent of the market for PC operating systems. Other technology companies did the obvious thing: they developed most of their software to be compatible with

Windows. Microsoft, much to the credit of its senior executives, right from the start actively *enabled* this kind of innovation, in which other companies aimed for extensive interoperability with Windows. In turn, the availability of a very broad range of affordable software applications—including word processing software, spreadsheets, educational software, games, and the like—secured and solidified Microsoft's dominant position in PC operating systems due to what economists call network effects.

The shift away from stand-alone PCs and toward server-centric networks in the late 1990s started to threaten Microsoft's enormous profits. In the paradigm of "client-server networks," users can communicate via "clients" such as laptops, but also via smartphones like the iPhone or BlackBerry. Powerful servers, operating out of sight of these end users, perform much of the heavy lifting in terms of actual computing. In essence, servers enable the sharing of computing resources among multiple users. Examples of this approach to interop are plentiful in the office environment: we share printers and store and share our documents, spreadsheets, and portable document format materials (PDFs) in networked folders. Servers, not unlike PCs, can also run software applications that are available to the users of the networks—the "clients"—on an ad hoc, on-demand basis. The server's operating system, which is a dedicated system, controls all these tasks.

The growing importance of servers (which now can even deliver software applications to PCs) posed a challenge to Microsoft's original and highly profitable PC-based model. Once servers began to support a similar variety of software applications as PCs, powerful server-based operating systems might well become an alternative to PC operating systems. This vision (or specter, from Microsoft's perspective) of a broad range of server applications was made darker by the fact that server operating systems are typically based on open standards and programming interfaces. These interfaces allow technology companies to develop applications without asking for permission or paying expensive licensing fees. The effect of these changes for Microsoft has been that it cannot control as much of the marketplace through its strong position in the operating system market. The value of that position is still extraordinary to Microsoft as a source of revenue, but it is less important as a way to control related markets than it once was.

Microsoft's leadership had reason to fear that server platforms could eventually destroy the need for a sophisticated operating system like Windows. The Windows operating system has to be updated every couple of years, making it the source of a highly reliable, long-term revenue stream, much like an annuity that generates a steady stream of cash for the software provider over time. For this reason, Microsoft did not want to share much information about how it worked or to let competitors, like Sun, get too close to it with their technologies.

The investigation by European competition authorities revealed Microsoft's sophisticated strategy to protect its cash cow. The software giant leveraged its monopoly power from the market for PC operating systems (the primary market) into the market for a particular type of server operating system, called work group servers (the secondary market). It did so by deliberately *restricting the interoperability* between Windows PCs and non-Microsoft work group server systems. What made the reduction of interoperability such a powerful shield against competition? In order to run a network of computers effectively, the server operating system has to work seamlessly with the operating systems of the PCs connected to the network—a world almost entirely dominated by Microsoft. Because Microsoft was in control of the interfaces that control access to the PCs' operating systems, it could exclude potential competitors in the secondary market by refusing access to the PCs' core functionalities.

The Microsoft case illustrates one of the key characteristics of interoperability: it is not an all-or-nothing feature. One of the most important disputes during the European Commission's investigation was whether or not Microsoft had provided enough interoperability, given the origin of some of the underlying protocols. In several important areas—such as authentication and authorization or directory services, for instance—Windows was built based on protocols that were originally open protocols available to the public and were also used in open standard computer platforms.

However, over time, Microsoft added extensions to those protocols, extensions that were protected by intellectual property rights and that restricted access to Microsoft-compatible products. Microsoft argued that a lot of information was still available allowing competitors to build compatible products. According to Microsoft, competitors could achieve compatibility by installing software that could translate, or bridge, among the different systems, or by reverse engineering, an elaborate process in which programmers take apart a piece of software and analyze its functionality.

The European enforcement authorities engaged in a massive fact-finding effort to sort out whether the systems were in fact interoperable enough and what approaches Microsoft had used to protect its market position. Their research included customer surveys and expert testimonies to find out whether or not Microsoft provided appropriate levels of interoperability. The commission found that the proposed work-arounds were extremely hard to find in practice and did not work out well for users—not exactly "consumer empowerment." Furthermore, Microsoft could easily just update its software and push the changes to their customers at any time. The effect would be to render useless all the work by a rival firm to make the systems interoperable.

The European authorities concluded that this level of interoperability was inadequate. Moreover, they ruled, Microsoft had deliberately reduced the interoperability between its Windows operating system and non-Microsoft server systems to a degree that effectively hampered competition by hindering market entry for other server operating systems. Against this backdrop, the European Commission—later backed up by the European Court of First Instance—ordered Microsoft to disclose complete and accurate interface information that would allow non-Microsoft work group servers to achieve full interoperability with Windows PCs and servers. Microsoft was allowed to charge a reasonable fee for the interoperability information it had to disclose. The court also gave Microsoft significant leeway in how it disclosed the requested information.

Given what was at stake, it is hardly surprising that Microsoft worked so hard to avoid disclosing interoperability information that would open the door for its competitors. Microsoft's lawyers argued, among other things, that the mandated disclosure would have a strongly negative impact on innovation by hampering Microsoft's incentive to create new software products. According to Microsoft, the mandatory disclosure of interoperability information would allow competitors essentially to copy software that Microsoft had developed and that intellectual property laws protected.

The European authorities rejected this line of reasoning. First, Microsoft was only asked to disclose "gateway" information, not the source code of the operating system. This disclosure of gateway information empowers competitors to develop their *own* interoperable software, but it does not give them a free pass to simply copy and resell Microsoft's products. Second, the competition authorities concluded that, on balance, the positive impact on the level of innovation of the entire industry outweighed any possible negative effect on Microsoft's incentive to innovate. In essence, the competition authorities found that more interoperability would help level the playing field by widening competition across the board. This, in turn, would increase, not decrease, innovation through diversity.

Although Microsoft's two competition-related cases in the United States and in Europe are the most prominent fights, it is far from the only technology company that has faced legal battles about the disclosure of interoperability information. In 2006, the French National Assembly, for instance, passed a law that regulated Apple's ability to offer iTunes in France as a noninteroperable music download system. Observers in the press dubbed it the "iPod law." The French legislators made an argument along lines very similar to those of the European Commission's: Apple needed to open up its systems to allow for greater consumer empowerment and greater competition in the market for digital music. The French regulators have yet to force Apple to make its system more interoperable, but the threat remains.[1]

The two Microsoft competition cases and the "iPod law" in France highlight the complicated interplay of interoperability and competition. Dominant market players often see it as in their interests to limit the amount of interoperability, whereas challengers almost always see interop as likely to help them, and the market at large, to develop. States see it as part of their job to keep the playing field level for competitors, at a minimum, and sometimes to help ensure consumer choice in the marketplace for information technologies.

The right answer to the question of how to promote competition through interop across the board is far from clear. It is hard to determine what level of interoperability is optimal for competition. It is just as hard to determine how much interoperability should be mandated when companies resist opening up on their own, without state intervention.

The lessons learned from cases like Microsoft's operating system and Apple's iTunes are shaping the policy framework for the digital age. In both instances, the companies have ended up making their systems more interoperable. Microsoft, in particular, has embraced interoperability as a major policy initiative across the company. Apple has been much slower to pursue interoperability across platforms, though it has made its own products extremely interoperable. The iTunes platform has grown more and more interoperable. These outcomes may have as much to do with consumer pressure for interoperable systems and with different corporate leadership as they do with legislators in Paris or regulators in Washington or Brussels. Regardless of the cause of the changes in Microsoft and Apple policy, the net outcome is both pro-interoperability and pro-competition, on a global basis, in both cases.

Although higher levels of interoperability tend to lead to more competition, it is important to note that there can be competition where there is no interoperability. Put another way, there can be high levels of competition among noninteroperable systems, too. The history of information and communication technologies is full of *competing* noninteroperable systems: VHS and Betamax videocassette recorders (VCRs), Macs and IBM PCs, 5¼ inch and 3½ inch floppy disks, analog television and high-definition television (HDTV), Blu-ray disks and HD-DVDs, Apple's iTunes (not so interoperable) and Microsoft's Zune (more interoperable)—these are only a few out of a very long list of examples. We need to ask hard questions about what happens if we leave such noninteroperable systems to compete against each other in the free market. And we need to focus on how consumers will be affected by these interoperability-related decisions. Before we ask and answer these questions, we need to establish a better sense of the powerful forces at play in today's high-tech markets.

In the last two decades in particular, we have witnessed the spread of high-tech products and services that exhibit *network effects*. Network effects are an important characteristic of most of the markets we address in this book and are central to an understanding of interoperability and why it matters.

A not-so-high-tech product can help explain what network effects are and why they are so important to our story of interoperability and competition: the old-fashioned fax machine. Imagine a hypothetical situation, a few decades ago, in which only a handful of organizations and individuals own a fax machine. Only a few of these groups actually have interoperable fax machines that are able to send and receive messages to and from each other. Intuitively, we know that the value any user gets out of each fax machine is relatively small. After all, any given user can communicate with only a very small group, the owners of compatible fax machines. Contrast this situation with an alternative scenario in which, say, millions of organizations and individuals around the world are using fax machines that are compatible and can talk with one another. Here, the fax machine as a sender and receiver in a communication network is of much higher value to each user. This phenomenon of the value of each product in a network increasing when compatible products are added to the network is a network effect.

In theoretical terms, a network effect occurs where the consumption benefit of a network good is proportional to the total number of consumers who purchased that good. Instead of a fax machine, one might think of a telephone (an example of a network effect taking hold before fax machines) or Cisco's TelePresence systems (an example of a network effect yet to occur).

Network effects are essential to the health and continued growth of today's high-tech markets. When multiple people communicate with one another via e-mail or cell phone, the consumption benefits for everyone on the network increase if the number of consumers on the same network increases. The same is true with document formats. When groups share files in the same file format (such as Word files or Adobe's PDF), the benefits to everyone in that network using those technologies go up.

Network effects can also arise in "virtual" high-tech systems that consist of a hardware component and (interoperable) software. These hybrid systems require the user to buy at least two components. Computer hardware and software, a TV set and video programming, or tablet PCs and downloadable applications are a few examples. The network effects in these examples are by and large the same as in the case of the fax machine. The value of an iPad for a user increases as the number of interoperable applications available for use on it grows.

Network effects have several interesting implications for competition in information technology–related markets. For example, network effects can have a big effect on the way consumers act. Network effects can shape how users decide among noninteroperable devices, software, and services. Counterintuitively, network effects can push markets toward a single technology standard, as the value—and often the ubiquity—of a particular product or service increases as more people use it. The popularity of the Windows operating system is a prime example of this phenomenon.

So why exactly is it that many people in network markets tends to use the same system despite the availability of numerous competing noninteroperable technologies? In many cases, consumers will choose a technology based upon what is considered cool, what their friends use, or the cultlike status of the company offering it. This mode of decision making can favor the first mover, the company that broke open the market and created the first sensation. The market for digital music players, dominated by Apple's iPod, is a good example of this phenomenon. A buyer of a portable digital media player, for instance, can choose from a number of systems that are (largely) not interoperable with one another: Apple's iPod, Microsoft's Zune, and Creative Technology's ZEN. Each system consists of hardware (the player itself) and software. The software in this case includes the songs, videos, or podcasts that the user buys, typically from an online store, and that is usually tailored for one particular make of player. Her product selection will depend on her preferences regarding the design, features, performance, and other aspects of the player. Her selection will also depend on her expectations about the availability, price, and quality of the other components. Put another way, because switching between, say, an iPod and a Zune is costly, the media content she will be buying in the future will affect her choice in the present. In this case, the user chooses the digital media player she perceives to be the most popular.

At this point, the vicious cycle of network effects can kick in. The non-interoperable system that seems to be more popular will actually become more popular for that very reason. Moreover, because the value of the system increases as new users are added, the value of the network increases for future adopters, which will make it more attractive to them. This dynamic plays out in such a way that even a modest existing user base, a *slightly* better reputation, or some brand-name recognition can influence consumer expectations regarding a particular technology. Once consumers make up their minds based on these factors and perceptions, it is very difficult for a competitor to catch up to an incumbent later in the game. Once a system has gained some initial popularity, it can pull away from its competitors and become a de facto standard. This phenomenon is called the tipping of networked markets.

This phenomenon has been responsible, for instance, for the triumph of VHS over Betamax videocassette recorders, even though many people believed Betamax to be the superior format. The competition between them is a modern classic in the business literature, in which JVC's VHS beat Sony's Betamax format through better marketing and network effects. When consumers flock to a well-marketed early product offering, an inferior technology can defeat a better technology through the magic of network effects. VHS and Betamax were not interoperable with one another; consumers had to choose between the two to play a rented movie or to record a television show. JVC's effective marketing of the VHS format led to its swift, broad adoption in the marketplace, which made it hard for

Sony's Betamax to gain ground. Once the videocassettes that people wanted to watch were in the VHS format, and once people bought the videocassette recorders to play VHS-formatted material, the network effect kicked in. Sony's Betamax never had a chance after that. Many video enthusiasts continue to believe that Betamax was the superior format on technical grounds and for ease of use. A few diehards continued to use the Sony format, but the vast majority of the market share tipped to JVC—until the DVD and other digital formats came along to supplant them both, of course.

Superior marketing does not always lead to network effects kicking in. Competition among noninteroperable goods in network markets does not always lead to de facto standardization around one single system, at least not in the short run. The variety of high-tech systems that serve similar needs on today's marketplace is ample evidence of this. Consider videogame consoles. Sony's PlayStation, Microsoft's Xbox, and Nintendo's Wii are in fierce competition with each other. Smartphones with different underlying operating systems—such as Apple iOS, Google Android, Windows Phone, Palm's WebOS, and BlackBerry OS—are another illustration of the same phenomenon. In cases where consumers are heterogeneous and products have distinct features, multiple noninteroperable systems can survive. In these cases, companies can also target consumers who value the features of one noninteroperable system more than the larger network size of the competing system. Apple's entertainment product strategy is illustrative in this respect. An iPad, for instance, does not offer interoperability across app stores. But the design and lifestyle built around the device is unique and compensates for the lack of interoperability.

The availability of competing noninteroperable systems—such as in video game consoles, smartphone operating systems, or tablet computers—comes with benefits for users who have preferences that are different from those of the majority. But society at large can also benefit from situations in which consumers can choose among competing systems and do not have to rely on one single technology or uniform standard. Paradoxically, benefits such as increased productivity or gains in efficiency are easiest to see by reversing the mirror. Consider, for instance, circumstances in which market forces settle on one particular technology standard that is less good from an engineering perspective than an available alternative (which, according to some, happened in the VHS and Betamax story). Economists call this phenomenon suboptimal standardization, and it helps explain why we as consumers sometimes have the impression that we live with 1.0 devices or services when we see 2.0 versions all around us.

A familiar example of such a technological lock-in is the QWERTY keyboard layout. QWERTY is the de facto standard for the English-language keyboard layout. It emerged as a standard based on market forces in general and on network effects (tipping) in particular. Christopher Latham Sholes, one of the inventors of the typewriter, was responsible for the QWERTY keyboard layout. Sholes came up with the layout in the late 1860s in order to solve an annoying problem that plagued early typewriters: the typebars frequently jammed when a user hit certain combinations of keys in rapid succession. Sholes's solution was to create a keyboard that minimized the frequency of typebar jams by placing the keys most likely to be struck in close succession far apart from each other.

Soon after the QWERTY typewriter hit the market, however, mechanical advances reduced most typebar jamming. From a technological viewpoint, the QWERTY design was obsolete soon after it was introduced. Alternative keyboard layouts have emerged over the years, but none has caught on. The most widely known alternatives are the Dvorak Simplified Keyboard and the Colemak keyboard layout, which is partly based on the QWERTY layout. Studies have shown that these alternative keyboards allow for faster typing than the original QWERTY layout. However, none of the superior layouts ended up replacing QWERTY as the de facto standard. Again, this technological lock-in has to do with network effects (think of the keyboard as hardware, and the experience on that keyboard as software). By the time better solutions entered the market, QWERTY was already widely used. The switching costs for learning a non-QWERTY layout keyboard are high for any individual user. The path of least resistance is to stick with QWERTY, which makes it very easy, for instance, to use keyboards on a machine at home or in the office, to work with equipment provided by a new employers, or to use a rental if the laptop breaks. As a result, QWERTY remains the standard to this day—well into a digital age, well after typewriters themselves have faded into history museums and dusty attics, and when the jamming of hammers as an impediment to typing is hard to imagine.

Network effects help us understand why some technologies succeed where others fail. From the perspective of our study of interoperability, the key insights are that competition can exist among noninteroperable systems, just as it can among interoperable systems. Competition among non-interoperable systems can also lead to the success of suboptimal systems, as the VHS-Betamax example shows. And whether the successful technology wins out among interoperable or noninteroperable systems, the problem of lock-in can arise in any scenario, which in turn is bad for innovation.

Network effects also help explain why companies make the choices they do about interoperability strategy. This decision-making process within companies can have a big impact on the degree of competition within a given information technology marketplace. Companies have to make choices about their interoperability strategy all the time. In designing a new product or service, much may turn on whether it is meant to interoperate with the products or services of competitors or not. Network effects are essential to the decision making within companies about whether to choose one approach over another.

Whenever a company believes it has an initial competitive advantage that can translate into a large and lasting one because of network effects, it is more likely to select a noninteroperability approach. But it is hard to know when a company has reason to believe that it will be the winner that "takes it all" by tipping of the market in favor of its product or service. Much also turns on the maturity of the company and of the market in question.

There are a number of possible ways to guess which approach a company will take when it comes to interoperability strategy. A company with a large, loyal user base has an initial advantage against its rivals, especially against competitors who are just entering the market. Because of its strong and growing user base, for instance, competing successfully against Facebook, the dominant social networking site, is very difficult—even for a company with the firepower of Google, as the recent launch of Google+ demonstrates. Thus, it would be a surprise if Facebook were to decide today to become perfectly interoperable with competing services. A strong brand and a great reputation are factors that can tip the balance in the market and create incentives for a noninteroperability strategy.[2] Apple's products with a strong "hipness" factor, such as the iPod or the iPad, are again useful examples. Product differentiation may also be a factor. For example, if a subset of users harbors strong preferences for a particular component of a product or service, a component that is unique to that product or service, the firm producing it may also have a reason to follow a noninteroperability approach.

A company's interoperability strategy is highly likely to change over time. Microsoft is again a constructive example. Microsoft relied upon the logic of network effects to leverage a relatively small initial advantage in the form of an installed user base to build a very profitable monopoly in PC operating systems. That is, Microsoft began with an approach based on moderate levels of interoperability but later switched strategy and took deliberate steps to reduce interoperability. But in the past few years, Microsoft has again changed its position, partly in response to the antitrust actions taken by governments on both sides of the Atlantic, partly in response to changing technological and market environments, and partly because of new leadership. Recently, it has made interoperability a major part of its company strategy, in the context of even its highest-profile initiatives, such as its push into cloud computing.

Microsoft has committed to implementing a range of important industry standards and to documenting that implementation in order to make it available to all software developers. One example is the Office product suite, released in 2007 and again 2011, which enables users to save and open documents in a variety of industry standards, including standards that Microsoft's competitors originally sponsored. In addition, Microsoft disclosed a great deal of technical information regarding the protocols embedded in its products that are used to exchange data with other products. For protocols that did not include patented innovations, the technical documentation was made available for free. In other cases, Microsoft committed to reasonable and nondiscriminatory terms for the licensing of its intellectual property.

Microsoft's current approach to interoperability is substantially different from its approach in 1992. Whether or not prompted to do so by regulation, companies change their posture toward interoperability over time. This decision often depends on a mix of market factors, a company's strategic approach to differentiation from its competitors, and the maturity of its services and products.

An increase in competition and diversity in the marketplace is one reason to seek interoperability. At the same time, the wrong level of interoperability can lead to a related negative effect: uniformity. Interoperability may lead, for instance, to the homogeneity of services in a given sector. Too much interoperability may lead to a world where all systems are so standardized that they become uniform and squeeze out variety or diversity—or where there is only one player addressing certain important complex problems.

The problem of homogeneity, however, is not an argument against interoperability; rather, it is a concern that emerges from the way interoperability is often achieved: through standardization. There are a range of different standards and standards-setting processes. Whether we look at Internet standards or air traffic control, the result of standardization is some level of homogeneity. In some sense, the reduction of diversity is the raison d'être of standards. More often than not, this harmonizing effect is actually desirable. It is helpful, for instance, to have one standard number (9-1-1) for placing an emergency call, whether we're in New York or Tennessee. Likewise, it makes sense to agree on industry-wide drug safety standards that apply across nations. It makes sense to ensure the availability of standardized CD/DVD drives in our laptops.

Intentionally or not, setting standards to ensure interoperability can have a leveling effect. This can become a problem if it leads to homogeneity, especially when circumstances change radically. A single platform, for instance, with which many systems interact becomes a de facto standard over a long period of time. The lock-in caused by many firms building upon a single standard might make it very hard to incorporate technological advances as they occur. In such situations, the effect of interop is not extensive innovation, but innovation constrained by what is possible on or within that platform. Air traffic control systems are a good example. Those who seek to reform the air traffic control system lament the difficulty of upgrading to a new system worldwide, despite the obvious long-term cost and safety benefits of doing so. One of the reasons it is so hard to reform air traffic control systems is the deeply rooted interoperability of the current system. A winner-takes-all system, in which the winner of an early standards battle becomes entrenched through strong network effects and many interdependencies, can make a workable but suboptimal system very hard to replace, even when there are good reasons to do so.

It is not interoperability per se that may lead to homogeneity. Rather, homogeneity might be a consequence (even unintended) of the means by which parties aim for higher levels of interoperability. This design problem is a close cousin of the argument we advanced in the context of security and privacy. The general idea behind interoperability—and indeed, what we argue is its beauty—is that it does not require all systems, applications, or components to be uniform.

Most of the time, a minimally invasive design approach focusing on interfaces or membranes can enable systems—whether technical, human, or institutional—to work together while still preserving their difference and variety. Only in certain (but admittedly important) instances does interop as a design matter actually require uniformity of the system overall—or lead to it unintentionally. And even in such instances, uniformity at one level is likely to lead to more innovation and diversity on other levels in other parts of the network. Take the Internet itself as an example. At the Internet's logical layer (the code and standards that make it run so seamlessly), a set of uniform protocols (called by abbreviations such as TCP /IP) are dominant. Despite the standardization at this layer of code, the things we can do on top of it are wildly diverse: we can listen to any kind of music, read any kind of text, build any kind of new application, and so forth. The system has standards at one layer (homogeneity) and diversity in the ways that ordinary people care about (heterogeneity).

Recall that there are different ways to bring about higher levels of interoperability. Standardization is only one approach among others. Even standards approaches do not automatically eliminate differences or lead automatically to uniformity, as many case studies from the consumer electronics world illustrate (take, for instance, the diversity of cell phone models, as a simple case). Nor are standards the only pressure point toward uniformity, as the example of fashion demonstrates. Many people like the same clothing style or the same small number of car models (consider the prevalence of the Toyota Prius in left-leaning communities today), even though standards in the interop sense play no role in those preferences.[3]

Interoperability strategies should address ways to avoid squelching the kinds of diversity that are productive. Diversity, as a concept and a value, is under threat in many arenas—from the natural world to the world of technology.[4] A true interop approach generally fosters and preserves diversity. The idea of interop is to embrace certain kinds of diversity not by making systems, applications, and components *the same* but by enabling them to work together. The designers of interoperable systems should develop granular design principles within the interop framework, principles such as, "While working toward higher levels of interoperability, harmonize systems, applications, and components at the edges and to the extent necessary— but no more."

It is one thing to acknowledge that more interoperability is good for competition. It is a much more complicated, and more context specific, task to figure out what levels of interoperability have the maximum impact on competition, how these levels can be achieved, how much progress toward interoperability can be left to the market, and when government intervention is appropriate. The corporate history of market leaders like Microsoft, which has changed its interoperability strategy over time, can be an instructive guide in understanding the complex relationship between interoperability and competition. A range of factors—from consumer expectations to state intervention to theories of market development—can lead companies to make different decisions about how interoperable to make their products and services.

Our view is that markets for information technologies tend to become more interoperable over time, as the cases of operating systems and digital music show. That tendency is a good thing for competition. But the process is rarely straightforward or linear. The equally complex relationship between interoperability and innovation, the subject of the next chapter, is a closely related problem.

CHAPTER SIX

Innovation

I nteroperability is an especially powerful tool for fostering innovation. Increased levels of interoperability at the right layer in a stack of technologies can lead to innovation at multiple levels. In the digital age, increased technical interoperability typically enables innovation at the human and institutional levels. As an example take Google Maps, a service that provides a basic infrastructure for geolocation information, upon which applications as diverse as restaurant guides or coordinated disaster relief efforts have been built. In theoretical terms, this quality is known as the *generativity* of the Internet.[1]

In other instances, interop-based innovations allow societies to harness the creative spirit of individual citizens. The diversity and creativity of user-generated content shared over platforms such as YouTube or Wikipedia are two impressive illustrations of the enormous creative power increased interoperability among digital devices, applications, and components can help bring about.

Interoperability also can (but does not always) help ensure that we do not lock in substandard technologies. In this way, interoperability does not just foster innovation directly; rather, it can help lead innovative technologies on the market to become more broadly adopted. This is a particularly important feature of interoperability in the context of innovation. Once a particular technology—such as a computer operating system—has become popular in markets with strong network effects, it is usually very "sticky" and hard to replace, even if a more innovative product or service arrives on the market. The power of interoperability as a means to overcome technological lock-in has been well studied by economists, but it has gained little attention from policy makers.

At the same time, the highest possible level of interoperability does not always advance the goal of promoting innovation. We have studied cases in which interoperability may have a limited or even negative effect on innovation under certain conditions. This is especially the case in situations where companies have strong incentives to innovate because they compete for the entire market instead of just a share of it. Apple's iTunes is an example in this category: Apple created a highly innovative, low-interoperability product that they saw as able to take an entire market.

Even in cases in which high levels of interoperability do lead to innovation in a given market, there is no guarantee that this positive, symbiotic relationship will continue. Interoperability has led to great innovation in the social web, for instance. When companies like Twitter and Facebook open up their APIs to others, innovators can hook in and build their innovations upon the open systems made available by a series of private firms. But problems may arise over time. One or more of the web services providers may decide to pull a bait and switch by introducing a fee for the kinds of connections they initially made open. In turn, the innovative services built on the highly interoperable systems of today may be cut off when companies seek to profit from their central place in the ecosystem. This problem may occur even if companies never seek to charge for interconnection; a company might, for example, go out of business, yanking a key building block out of a complex system. The point is simply that what works in favor of innovation on day one may not work the same way later.

We revere innovation. It is a central goal of public policy at nearly every level. We have embedded the concept of innovation in the US Constitution, in key pieces of legislation and regulation, in court decisions, and in policy statements. Innovation is an official public policy goal of every modern society and of every society striving to modernize itself and to grow. We turn to innovation to help us solve the massive societal challenges we face today—ranging from global warming to the health care crisis. We all have to think hard about how to work together to promote and support innovation.

A major earthquake struck Haiti a few hours ago. Is there any way for us to help? Thanks, Patrick.

—*January 12, 2010*

A terrible earthquake hit Haiti on January 12, 2010. In its wake, it left somewhere between 92,000 and 220,000 casualties, and around 1.5 million to 1.8 million Haitians homeless. With this brief message, Patrick Meier, a graduate student at the Fletcher School of Law and Diplomacy at Tufts University, reached out to a group of about three hundred volunteers.

Meier and his friends decided there was a way they could help. They worked together to launch a Haiti disaster-relief effort. This lightly structured, instantly formed association brought together Ushahidi (a nonprofit technology company that develops software for information collection, visualization, and interactive mapping), for whom Meier was working as director of crisis mapping; the Fletcher School; the United Nations; and the International Network of Crisis Mappers. Within a few hours, hundreds of humanitarian and technology workers who had not previously known one another signed on to join the start-up initiative.

The Ushahidi platform facilitates large-scale collaboration among disparate, and otherwise uncoordinated, users by allowing them to share information about events and crises in real time.[2] A group of activists developed the original website in the aftermath of Kenya's disputed 2007 election. Across the country, eyewitnesses to both violent acts and graceful efforts to support peace submitted information via web or text messages to the new Ushahidi platform. Ushahidi provided a simple, powerful way for observers to record and situate these incidents on Google Maps.

Ushahidi quickly grew to have over 45,000 users in Kenya alone—despite the country's relatively low level of overall Internet penetration. Ushahidi established an online community of citizen journalists, activists, and ordinary people, most of whom did not know one another beforehand, in a virtual network. Soon, similar sites began to spring up on the platform, focusing on other regions and purposes, including the tracking of anti-immigration violence in South Africa, violence in the eastern Congo, and depleted pharmacy stocks in East Africa.

The Ushahidi model soon extended far beyond Africa, with similar sites being developed to monitor elections in Mexico and India and to collect eyewitness statements during the Gaza War. The platform was used in Russia to set up a map to help volunteer workers during a series of terrible wildfires in 2010. Many of the volunteers worked from countries thousands of miles away to help coordinate the firefighting on the ground.

Interop drives innovation through the Ushahidi platform by letting people create highly interconnected systems on the fly in moments of crisis. The group of young and very engaged activists developed Ushahidi at a low cost and very quickly, and they made it available to the world as a platform for innovation in the public interest. The platform is a particularly creative manifestation of what we have discussed under the rubric of consumer empowerment. Ushahidi has interoperability written all over its DNA. It is grounded in the principle of open source. As a mashup, it combines a number of existing components and elements in innovative ways.

The power of the Ushahidi model derives from the way it establishes and maintains, on behalf of its distributed users, high levels interoperability among a series of devices and data formats. Ushahidi connects different devices, such as computers and mobile phones, and is designed so that it can receive messages regarding events via Short Messaging Service (SMS), e-mail, or tweet or through the website itself. Users tag the reports and then locate them on a map. Google Maps API performs the necessary geocoding.[3] The interface can be built off of either Google Maps or OpenStreetMap, a collaborative project to create a free and editable map of the world. In addition to text reports, a user can also submit photos and videos that can be integrated into the maps.

Ushahidi's technical features are not the only reasons for its phenomenal success in the Haiti disaster relief effort and for its use in many other crises around the world. Perhaps even more notable than the mashup itself is the fact that its technical interoperability enables innovation at the upper layers of our interoperability framework, including the institutional layer. For example, Ushahidi provides a single information source for individuals and organizations who are on the ground during a crisis and want to work together.

By submitting reports of particular events, which are then tagged and mapped, these disparate operations can organize and coordinate. Various organizations have access to this map and can use it to target rescue efforts, to investigate violence, or to engage in other activities that are made easier and more efficient by facilitated coordination within and among groups. After a crisis has ended, other organizations can make use of data collected over Ushahidi. Chronologically geotagged and verified information, for instance, allows researchers, historians, courts, political movements, and nongovernmental organizations (NGOs) to gain a better understanding of how a certain critical situation has emerged. More profoundly, the knowledge generated by a broad community through Ushahidi informs the way we might develop early warning systems in the future.

The technical and data interoperability harnessed by Ushahidi can in turn generate interoperability at the human and institutional layers. The system enables after-action review and analysis of the initial events and the institutional responses in real time. A review of the Haiti crisis information-management efforts, for instance, graphically demonstrated the need for more coordination within the crisis-mapping community. It also underscored the importance of interoperable processes and cultures across the different constituencies.

Higher levels of interoperability at the human and institutional layers is already leading to next-generation crisis-mapping applications, risk-prevention and risk-reduction programs, and long-term recovery processes. But interoperability does not always drive innovation at these layers, either. The caveat that we explored in Chapter 5 on the effect of interoperability on competition applies here as well: too high a level of interop can also lead to uniformity and lock-in, which can work at cross-purposes to innovation. Despite this potential downside, the overall relationship between innovation and interoperability tends to hold across all four of the layers of interoperability. This relationship helps explain the rapid development not only of specific services like Ushahidi but also of the web itself.

I nnovation in web services has been central to the evolution of the web over the past decade. This innovation derives in large part from the availability of different data sources and functionalities obtained via multiple open APIs.[4] Interoperable web services that are mixed and mashed up allow different types of innovation to occur on top of the technology layer. This approach to web development enables innovation to spring from unexpected places: from unanticipated combinations of existing data, creation of new content by analyzing existing data, the evolution of new business models, and many other forward-looking approaches. Mashups illustrate the key point of this chapter: optimal levels of interoperability in digital environments can foster substantial levels of innovation.

Open APIs allow anybody—professional developers and geeky amateurs alike—to access the data or services of a platform. As one of the key ingredients of web mashups, open APIs are massively powerful drivers of innovation. Moreover, nonprogrammers can gain access to a malleable form of the data. Among the most popular APIs are those associated with Google Maps, Flickr, YouTube, Twitter, Amazon eCommerce, and Facebook. Each of these APIs is asked to transfer data across systems (requests known as API calls) billions of times per day.[5] Many of the companies behind these web services have discovered and benefited from the enormous potential for innovation that is unleashed when the web community is encouraged to freely mix data and functionalities in cases where users are tackling a particular problem.

The extraordinarily rapid rise of Twitter demonstrates the relationship between interoperability and innovation. Twitter was created and launched in 2006. It was released as a microblogging service, and it enables users to send and read text-based posts of up to 140 characters ("tweets"). The founders sought to create a new way for users to share information in a concise way. Beyond this, they did not have any specific goal or purpose in mind for Twitter. Twitter gained popularity among early adopters in 2007 but left many wondering whether it would ever really take off. It did. As of February 2011, roughly 190 million users were generating about 65 million tweets daily. Moreover, Twitter has become one of the most visited websites in the world. Once viewed skeptically by some, Twitter now plays a significant role on the global stage—for instance, during the Arab spring in 2011, protesters in several countries relied heavily on Twitter to organize and publicize their cause.[6]

As of 2010, Twitter supported a staggering 70,000 applications—virtually none of them developed by the company itself—and the application base continues to grow.[7] The expansion of Twitter's reach was no doubt supported by the release of its API in September 2006, which effectively allowed many devices and web applications to interoperate with Twitter. The hope was that users on these media would create and disseminate information in new and innovative ways, and indeed, a broad range of Twitter-based applications rapidly emerged. Such applications include a news service for stock traders, an executive search service, tracking services for travel sites, and even a service that lets users submit, vote on, and create T-shirts from tweets.

Facebook is another powerful demonstration of how interoperability via open APIs can drive innovation. When Facebook released the first version of its API in May 2007, third-party developers created thousands of new applications within six months. Just as with Twitter, the applications developed to interoperate with Facebook cover a broad range of functionalities. These applications serve wildly diverse needs—from games and sports applications to business tools, utilities, and educational applications. Currently, the most successful apps on Facebook are games from a company called Zynga, which has roughly 297 million monthly active users. But new tools have been created not only by companies but also by young entrepreneurs. Scrabulous, a popular adaptation of the word game Scrabble for the Facebook platform, is a prominent example. Recently, Facebook introduced a new and improved API that makes it easier for developers to use Facebook as a platform for their own innovations.[8]

It is also possible for users who are not programmers to create mashups. For example, ZeeMaps allows users to create free, customized, interactive maps and to add to it markers that are submitted in an Excel spreadsheet or created via wiki by crowdsourcing.[9] Pipes is another powerful composition tool used to mash up content from the web. Among other things, it helps users aggregate, sort, filter, and translate feeds and locate and browse items on maps.

The power of interoperability extends beyond the consumer domain and into the business world as well. Mashups used in business settings—called enterprise mashups—illustrate another important, and closely related, phenomenon. Driven by business users who want easier access to enterprise data regardless of the application in which it is stored, mashup innovation also plays an increasing role in the enterprise context. Industry heavyweights such as Intel, Bank of America, Hewlett-Packard, and Adobe created an initiative called the Open Mashup Alliance to drive interoperability among business applications. Their approach was to promote usage of the interop-friendly Enterprise Mashup Markup Language. This alliance is another case of an industry-driven standardization process aimed at increasing interoperability among systems.

Facebook, Google, Bank of America, and Hewlett-Packard all have a strong interest in making data and functionality available to their customers and to business partners via open APIs. This is true even though they will not immediately capture all the value from the innovation unleashed through their platforms. Mashups enable companies to pull together disparate information and make it available in a form that is most valuable to customers. For instance, Facebook users may find it useful to see information they have posted to other services, such as foursquare or Twitter, appear in their news feed on Facebook as well without having to enter the information twice.

As a corollary, the more easily a programmer can customize data or functionality to serve a certain purpose for end users, the more we will observe the emergence of small, niche mashups. When individual developers modify existing web services for their own needs and then make the resulting mashup freely available to others, many more mashups will be made than would be the case if a significant investment of capital were required. Given that this development is so inexpensive and the developer does not have to get permission for every interconnection, modest advertising revenue can suffice to make a niche mashup profitable.[10] Often, however, mashups are not the primary revenue driver of a new business model. Instead, mashups facilitate or complement another business model, as the Facebook advertising model demonstrates. Facebook has developed a highly profitable business model based largely on advertising rather than on charging developers for interconnection to the service. As a result, the more developers build innovative ways to connect in to Facebook and the more people use the combined services, the more revenue Facebook stands to earn.

Outside the venture capital–funded start-up scene, nonprofits, governments, and private citizens also use mashups to serve the public interest. DataCalifornia is a service using the APIs of Facebook, Twitter, and Google Maps to view details and comments on California's education, health, and other current legislation. It also promotes collaboration by allowing users to submit ideas on how the government should spend or save taxpayer money. Congress111 is an iPhone app that mashes up a number of different Congress-related data sources; with this app, a user can view Congress-related news, votes, videos, tweets, and office maps. The Federal Communication Commission's consumer broadband test API provides up-to-date speed test data for wired and wireless connections.

Web mashups show how higher degrees of interoperability can be good for innovation. In the research for this book, we conducted a series of case-specific studies with our team. We investigated potential examples, such as identity-management systems, to confirm the relationship between increased levels of interoperability and innovation. In the identity-management business, firms are working to help reduce the number of times a consumer has to log on to one system or another. The common complaint "I can't remember all those usernames and passwords!" is a consequence of noninteroperability across systems. When a site allows users to log on using their Twitter, Facebook, or OpenID accounts, they do not have to waste as much time switching between services and experiences. The more systems agree to work together, by accepting common forms of identity management, the more innovation flourishes in identity management specifically and on the web in general.

Although these narratives provide powerful anecdotal evidence of the connection points between interoperability and innovation, it is much harder to glean why, in a general sense, this positive relationship so often exists. Interoperability theory offers two possible explanations. First, interoperability usually increases competition, which in turn is expected to lead to higher rates of innovation. Second, interoperability also tends to reduce the effect of lock-in and lowers the entry barrier for entrepreneurs. Take a look back at the Microsoft story in Chapter 5. Forcing the software giant to disclose information allowed existing rivals and new market entrants to compete by enabling them to build new—but interoperable— products and services. Such products and services not only permitted users to switch between providers, but they also allowed users more freedom to use applications that were running on top of them. Enhanced competition of this sort benefits users by reducing prices and by providing companies with incentives for product and service innovation.

A word of caution is necessary here. The interoperability-competition-innovation progression can sometimes get complicated. Some economists argue that interoperability can even have a negative effect on innovation by leading to anticompetitive situations. For instance, standards-setting agreements among companies can lead to more interoperability and more innovation in the short run. However, such arrangements may prompt a single firm or a few firms to act anticompetitively in the long run.

When a standards consortium manipulates the standards-setting process to achieve anticompetitive ends, a related problem can arise. In the

USB 2.0 standards-setting process, companies were working together in the lead-up to the year 2000 to come up with a new protocol for connecting peripherals—such as keyboards—to computers and for sharing data quickly among devices. At least one company is alleged to have used an information advantage for anticompetitive goals in the course of this standards process.[11] These are valid concerns; nonetheless, we argue that such anticompetitive actions reflect the unscrupulous behavior of a specific company rather than a flaw of interoperability itself.

The Microsoft case brings up a further complicating factor in the relationship among interoperability, competition, and innovation. In that case, Microsoft argued that the forced disclosure of interoperability information might result in decreased competition. In essence, Microsoft asserted that the disclosure would have the damaging effect of reducing its incentive to invest in the development of new products and services. More generally, firms may have a stronger incentive to be innovative when low levels of interoperability promise higher or even monopoly profits. This sort of competition (economists call it Schumpeterian competition after the famous Austrian economist Joseph Schumpeter) creates incentives for firms to come up with entirely new generations of technologies or business methods that are proprietary. Apple's iTunes strategy is a case in point of a company competing for the market as a whole rather than for only a share of it.

Despite these complications, there is broad consensus among economists and regulators (recall, for instance, the European Commission's response to Microsoft's argument) that competition is good for innovation at a *marketwide level*, even if not necessarily for an individual firm. Moreover, competition is just one of the theories that links interoperability and innovation. As the mashup example illustrates, innovation in the Internet age does not only happen as a result of the competition-driven activities of companies and their respective R & D labs. Rather, if the underlying platforms are open and designed with interoperability in mind, then end users, intermediaries, and other actors contribute in distributed and often vertical ways to the development of new products and services.

The power of openness and interoperability for innovation is among the most fascinating aspects of the Internet. In his much-acclaimed book *The*

Future of the Internet—And How to Stop It, Harvard professor Jonathan Zittrain put forth what he calls the "theory of generativity." By tracing the Internet's evolution and discussing its trajectory, he argues that ICT platforms should remain broadly open so that users can make creative developments on top of the ICT infrastructure;[12] the Internet would thus remain generative. Interoperability fosters openness of information and communication systems and is therefore a key enabler of generativity.

The powerful idea of horizontal innovation networks adds further heft to this line of argument about interoperability and innovation. Horizontal innovation networks are networks in which firms and users form porous, ad hoc teams to innovate. The work of Eric von Hippel, a professor at the Massachusetts Institute of Technology, highlights a key aspect of this idea: the importance of innovation for users who operate outside the traditional firm. According to von Hippel, two conditions are required to sustain innovation. First, at least some users must have sufficient incentive to innovate and to reveal their innovations. Second, the production and diffusion of these user-created innovations must be low cost and must be competitive with commercial production and distribution. Mashups are a great illustration of this type of user-driven innovation. In each of the successful mashups we have studied, a group of people with a common interest shared a desire to solve a problem and had a clear sense of what creative solutions were possible. Again, interoperability is one of the key enablers of this type of user-generated, low-cost innovation. Interoperability allowed like-minded people to work together, to experiment collectively to solve common problems, and to implement their ideas with limited expense.

The ability to make small changes—incremental innovation as opposed to radical innovation—is a third force that enables interoperability to foster innovation. Many new products and services are actually incremental improvements on an existing product or service. This "small-step" innovation builds largely on prior knowledge and resources. Technological advances associated with incremental innovations can appear rather modest initially, but their impact becomes profound over time. An example is SMS. The ability to send a short, text-based message may seem trivial, yet it has had an enormous effect on the way humans communicate in the early twenty-first century. (Think also of the short step between SMS and the 140-character tweets one can share, publicly or privately, on Twitter; of the advent of Twit-Pic, which allows the sharing of images on a similar network; and so on.) These advances occur much more frequently than advances arising out of radical innovations. Small-step innovation throws into sharp relief the role of interoperability in innovation generally: by increasing the level of inter-operability, more systems, components, or applications can be combined to make improvements on products and services. The range of potential improvements of the technology, in turn, grows broader over time.

T he promotion of progress and human welfare depends not only on the development of new technology but also on its *diffusion*.[13] Inter-operability also facilitates user adoption of high-tech innovations. The problem of adoption in high-tech markets goes back to the phenomenon of network effects. Consumer expectations regarding the future success of a new technology in a network market is a crucial factor in its success. This insight is relevant where consumers face choices with uncertain ramifica-tions. They can stick with a well-established, even if outdated, system or switch to the latest and hottest technology, which may never catch on.

There is plenty of empirical evidence that shows how consumers' ex-pectations about the availability, price, and quality of the components of high-tech systems shape whether or not they adopt a new technology. Con-sider the less-than-smooth transition from good-old analog television to high-definition television (HDTV). From the 1990s to the early 2000s, the FCC and Congress tried to shift broadcasting from analog to digital, with the hope that the transition would free up precious spectrum for other uses while enhancing the sound and image quality for consumers. Offi-cially, the digital television transition—often somewhat more dramatically called the "analog switch-off"—occurred in 2009. But the FCC and Con-gress had planned such a move since at least 1987. Consumers who were reluctant to make the leap were viewed as the primary holdup: why buy new, expensive HDTV sets when the variety, availability, and quality of HD broadcasting is uncertain? A brand-new TV is not of much value if there are few programs a consumer can watch on it. An economist with similar concerns might say that the utility of HDTV sets to consumers only increases when more HD broadcasting becomes available.

More problematic, broadcasters ensured that the availability of more HD broadcasting would be contingent upon increased HDTV set sales. It would not make much sense to produce TV programs using a cutting-edge technological standard that would be viewed by only a handful of households. Network effects give rise to what is essentially a chicken-and-egg problem. Rational consumers wait to adopt the new hardware (HDTV sets) until enough software (HD programming) is available. Conversely, software producers will probably delay investment in software (HD programming) until a critical mass of consumers have adopted the hardware (HDTV sets).

There is an additional wrinkle. Economists have observed that users tend to stick with an established technology even if they would benefit from switching to a new but incompatible technology. This wrinkle arises out of a mismatch of cost and benefit to current and future consumers. Consumers of an established product must bear the transition costs from the old to the new technology. All things being equal, they are less likely to switch than if they had not purchased the old technology in the first place. However, *future* consumers, who do not face this switching cost, would prefer widespread adoption of the new technology. As a result, markets for systems lock parties in to obsolete standards or old, suboptimal technologies.

The HDTV story is a good illustration of the problem. Consider a consumer who recently purchased an analog TV set. She currently receives all her favorite TV shows and movies, along with hundreds of channels. As it stands, she has little incentive to switch to digital TV—in terms of both the TV set itself and the availability of programming. At the same time, however, other consumers may be willing to buy an HDTV set because more programming has become available. The consumers who drive the adoption of a new technology often ignore the fact that some consumers will be stranded with the old technology.

In the case of HDTV in the United States, the government played an important role in overcoming these inherent problems by managing the transition from analog to digital TV. It used a multipronged approach to solve the problem, which included both an awareness and education campaign for consumers and legislative action.

Here, interoperability comes back into play. Among other measures, the government launched a program that provided households with coupons to buy a converter box that would enable them to receive digital signals, even on their analog TV sets. This was an inelegant yet effective way to create a modest level of interoperability between the old technology (analog TV sets) and the new (HDTV). Technical interoperability not only fosters innovation; it can also reduce the likelihood that consumers might be "stranded," or locked in to outdated systems.

I n the past three decades, the debate about how, precisely, societies can foster innovation has become intense. Some people believe that the answer lies in the structure and breadth of intellectual property (IP) regimes. They argue that we should increase the IP rights we award to creators in order to promote higher levels of technological innovation. Critics of the stronger-IP strategy, in contrast, warn that the expansion of IP rights is not the main catalyst for innovation. They argue that a stronger and broader IP regime will actually backfire by unduly raising the costs for future innovation.

There is broad consensus that the Internet is among the most innovative catalysts of our age. However, there is disagreement as to the appropriate means of fostering continued innovation of the Internet and related technologies. Proponents of an open Internet believe that an open and decentralized infrastructure maximizes its potential for innovation. Those opposed to this viewpoint see network owners, rather than users, as the primary motor of innovation. As a result, they favor a much more controlled Internet with strong property rights. Admittedly, each position has its merits and demerits.

For our part, we belong to a school of thought that believes in the innovative power of an open Internet. Our preference is for balanced systems of intellectual property protection that rewards creators while also recognizing the importance of the public domain. However, this debate, though still raging, is not the point of this particular book. We are focused here on the role of interoperability, as well as on those policies and practices that support interop in its best forms.

The key point we want to make here is that in these controversies about intellectual property policy, interoperability plays an important role. Our argument is that interoperability is one of the keys to innovation, especially in the case of information technologies (but, as we have seen, not limited to this sector). It is one of the key enablers of Internet generativity. Interoperability in this sense is important because it fosters the development of innovative technologies on top of the core technological infrastructure of the Internet itself. In turn, systems can function in much more efficient ways, to the benefit of individual firms, consumers, and society at large. Public policy ought to recognize and make explicit society's shared interest in accomplishing optimal levels of interoperability in order to foster higher levels of innovation.

The important role of interoperability for innovation should also be acknowledged more explicitly and considered more carefully in the heated debates about IP rights. The relation between IP rights and interoperability is complicated, and much depends on the specifics of the law. From a bird's-eye perspective, there is plenty of reason to believe that the current IP system is not designed in such a way that it will lead to optimum levels of interoperability in areas that are particularly relevant in the digital age. Take, for instance, the controversial case of patents protecting business methods or software, which can make it very hard and risky—and sometimes even outright impossible—for competing Internet companies to build interoperable services at the technology and data layers. Such strong intellectual property protections work at cross-purposes to interoperability and, thus to innovation in this case.

But it is not only the general and steadily expanding scope of what can be protected under today's IP regimes that tends to be bad for interoperability. To make things worse, lawmakers around the world, heavily influenced by the copyright lobby, have even enacted legal provisions that directly *prevent* the creation of interoperable services. Anti-circumvention laws have added a top layer of legal protection to the "digital locks," such as copy and access controls, aimed at securing copyrighted materials like music, movies, or e-books. In many countries, these regulations do not include exceptions that would allow competitors to open up the digital locks for the sake of interoperability.

We believe that specific laws preventing interoperability in the digital age are a bad idea, as they will have negative effects on innovation in the long run and thus should be changed or abandoned altogether. But IP law does not necessarily need to be in conflict with interoperability. Legal provisions, for instance, that carve out exemptions for reverse engineering of software for purposes of interoperability are a specific example in this category. More generally, IP regimes can be designed so that they make it easier for rights holders to enter licensing agreements—which is one approach to increasing interoperability as we have seen before.

The effect of IP rights on interoperability and innovation depends heavily on how those rights—regardless of their scope and shape—are actually *exercised* by rights holders. The use of Creative Commons (CC) licenses is illustrative in this respect. In the case of CC licenses, copyright enables the exchange of creative materials across systems, applications, and components. Creative Commons takes a permissive approach to IP and lowers the transaction costs of deal making, which in turn fosters interoperability at the data layer. We face not only the challenge of getting IP law right as a policy matter, but also the challenge of thinking more creatively and openly about how to wield IP rights on behalf of interop and innovation in the high-tech environment. That is the shared responsibility of lawmakers, company leaders, and consumers alike. The net result, if we are work together well, can be lower transaction costs and greater innovation across the board.

CHAPTER SEVEN

Systemic Efficiencies

Higher levels of interoperability can lead to systemic efficiencies. Although we often think in terms of the benefits to consumers—for example, from cell phone chargers working with more than one model or an app for Twitter working nicely with an app for Facebook—the greatest beneficiaries of interoperability are often business operations that use it to streamline their processes and manage costs. These businesses range from the smallest mom-and-pop stores to the world's largest banks and manufacturers. The flip side of systemic efficiencies is systemic complexity, which can lead to disastrous results if not managed well, as we have seen most vividly in the financial sector.

The role of interoperability in promoting systemic efficiencies can be observed by considering the most prosaic of business models: the grocery store. In the typical retail outlet of a few decades ago, employees used devices that looked like guns to stick small white price tags on each product in the store. A customer brought his cart to the front of the store, where a cashier would pick up each item, look at the price tag, and then key in the price by hand. A cash register, unconnected to any other device, would calculate the total.

This manual process came with many embedded costs to everyone involved, costs that could be stripped out with a little ingenuity. This manual process took time for both employees and customers. Worse still, the store's owner did not have an accurate idea of the sales trends for each product, except when she decided to call time out to take a department-wide inventory (by shutting down operations, perhaps on a monthly basis). When the retailer thought she needed to order new products, she would strike a deal with her suppliers, with only partial information on either side. In the negotiation, both the retailer and supplier would guard information about the inventories that they had on hand, as well as their sales histories and cost, so as to retain as much bargaining leverage as possible. When the order was delivered, an invoice would usually accompany it, informing the retailer at the moment of receipt whether certain products were unavailable or had to be substituted. In general, the method of these transactions required both the retailer and her suppliers to house inventory in back rooms or warehouses—a costly proposition—only to be drawn down as the retailer sold her goods.

This system, or a close variant of it, has been used for centuries. It is still used in many parts of the world. Over time, as improvements were made to the system of taking on inventory and selling goods at retail to consumers, changes tended to address a single aspect of the process to try to speed it up. The introduction of the cash register, for instance, facilitated a faster and more accurate checkout. Innovation in the automobile industry made trucks faster, which in turn, along with improvements to roads, increased the speed at which goods could be delivered. The basic system stayed the same, but as time passed, each piece came to be implemented more quickly and efficiently.

The introduction of a highly interoperable technology, bar codes, radically changed the way the grocery store, and many other businesses, operate. When products carry bar codes and retailers employ scanning devices at checkout, it is no longer necessary to price every product. Purchases can be made a lot faster today than in 1962, with the quick swipe of the product and a credit card.

From this perspective, the impact of implementing bar codes seems not unlike faster transport or the use of a cash register—each of these technological advances increases efficiency, speeding up the system in a marginal way, not radically. However, on closer analysis, the effect that bar codes have had is much more profound. Bar codes give computers a way to recognize individual items at checkout. This allows retailers to collect information on sales trends and makes it easily and immediately available (in contrast to the monthly inventories that they might otherwise rely on). Sales trends are very valuable data for both retailers and suppliers, given their interest in meeting consumer demands. With the introduction of bar codes, suddenly retailers and suppliers had a considerable incentive to work together and share information, activities that were further facilitated by the introduction of electronic data interchange (EDI), automated messages between businesses or organizations. The result: a radical shift in retailer–supplier relations. This process has culminated in a new, highly efficient strategy for managing the supply chain, or *just-in-time* manufacturing and inventory management.

Bar codes and EDI are examples of how greater interoperability can make a system substantially more efficient. In complex systems, the connection of previously unconnected parts can offer entirely new ways to improve the system. And even a connection that already exists can be made smoother and more productive. In this chapter we will first examine the benefits of higher degrees of efficiency and look at how interoperability can help get there; then, in Chapter 8, we will turn to the corresponding cost, which is radical complexity and its potential to cause extraordinary harm.

In order for our argument about interop's relationship to systemic efficiency to make sense, it is helpful to get abstract for a moment. Our argument is set in the frame of a basic term of economics: utility. Utility is a way of measuring whether one outcome is preferable to another. An outcome that produces one hundred utils is, strictly speaking, better than an outcome that produces ninety-nine utils. Note that measures of utility can only say "better" or "worse," not "good" or "bad." Any number of utils is meaningless on its own. It only has meaning relative to another measure.

Economists generally assume that, subject to some constraints, individuals maximize their utility. A system in which no one can capture more utility without making someone else worse off is what we mean when we describe something as economically efficient. The more efficient a system is, the closer it is to making everyone as well off as they can possibly be. We use utility as a measure of how efficient a system is.

Let us spend a moment in a purely theoretical place: the economist's ideal world. In this happy land, there are no mistakes or human errors. The manufacturers will never short the retailers, intentionally or otherwise, and the retailers know it. Thus, retailers do not need to spend anything to review the goods. As you might imagine, this makes things fairly simple. The retailers do not have to spend any time or money finding out exactly what they received, so they can simply order exactly what they want. Bingo: the system achieves maximum potential efficiency.

Of course, in the real world, things are bit murkier. Consider again the example of the common retail transaction. We know the manufacturer isn't perfect. Human error may lead to a staff person at the manufacturer's facility sending too few or too many of the goods that have been ordered by the retailer. (Set aside the possibility that the manufacturer is trying to rip off the retailer, which is another scenario altogether.) The retailer knows that the manufacturer periodically makes mistakes of this kind, so she makes the rational decision to verify that the goods she received are actually the ones she ordered. She hires workers to review incoming shipments, which means that receiving and inventorying goods costs her both time and money. Perhaps that process can be more or less perfected. The workers are walking briskly on the optimized routes. There is little or no additional utility to be captured by having the workers moving a bit faster. The system is as close to perfect as it can possibly be, right?

Not quite. Even if the existing solution to the inventory problem is implemented as well as possible, there may be a totally different and much better solution available. The implementation of the system being used and the relative efficiency of the system itself are two separate issues. When the system's implementation appears to be as good as possible and yet we still want the system to work more efficiently, change needs to happen at a higher level: at the level of the system itself.

Interoperability can be a partial solution to these types of inefficiencies. Interoperability offers surprising possibilities for fundamentally changing systems to render them more efficient. It can create new connections between previously distinct parts of the system, offering opportunities for smoother motion throughout the entire system. Once higher levels of interoperability are introduced, they bring with them both complexity and the opportunity for substantial growth in efficiency.

C omplex systems are never, in practice, optimally efficient. In the most complex systems, only part of the total possible utility in the system can be captured. Part of the reason is that there is ordinarily a collective action problem at work: firms and individual actors have a range of motives and incentives that are not perfectly compatible. Another reason, when it comes to the interoperability problem, is that it is impossible for humans, or even the computer models we put together, to comprehend the whole of the most complex systems. It may not be obvious that there is a way to improve the system—not without completely deconstructing and reconstructing it, anyway. The necessary new technology may not be available yet, the company may not have thought of places to make improvements, or partners in the supply chain may not be ready to work with the company in these new ways. But when a more efficient system emerges, it is almost always undeniable that all parties would become better off by adopting it.

In general, utility is a social good; we want to capture more of it, if at all possible. We will never be able to capture all the surplus of the most complex of systems. Put another way, there will always be some amount of deadweight loss. Our goal, from the viewpoint of simplified economics, is to minimize the deadweight loss built into the system.

One way to look at systems is to see them as made up of actions and activities. Each costs time, money, or both. The workers inventorying the goods cost both: time to complete the inventory and money to pay for their time. As intermediate costs, such as the total amount paid to the workforce, fall, the system becomes more efficient. The process of taking inventory would undeniably be more efficient if it could be done instantaneously and for free—or, better yet, if it did not need to be done in the first place. Neither of these is possible, but they do give goals to measure the real world against. Each real-world reduction in costs brings us closer to the ideal world.

Most costs affected by interoperability can be described as transaction costs. In the economics literature, "transaction costs" broadly means the costs to engage in an economic activity. For example, imagine buying a shirt on sale. You paid $50 for the shirt, but you also, probably without noticing, expended the time and energy to go to the store, hunt through the sale section, and try on shirts until you found one you liked. There is no way to eliminate these transaction costs, but they can be limited in a wide range of ways, often through the introduction of information systems.

Now let us apply the idea of transaction cost to a more complex system. Return again to our retail store owner receiving goods. She buys goods from the manufacturer. The manufacturer's transaction cost is shipping the goods to the retailer. The retailer's transaction cost is waiting for the goods to arrive and inventorying them when they do. A more efficient way of reviewing the goods, then, would reduce the retailer's transaction costs, allowing her to capture more utility.

Transaction costs do not have to be monetary, although they commonly are. They may be time costs or even more unusual costs. Such costs describe what happens when action happens within the system, whether the system is technological, a market, or otherwise. Unnecessary transaction costs make a system less efficient; they increase costs somewhere within the system, turning possible utility into deadweight loss.

Consider manufacturers and retailers. Retailers receiving boxes of widgets from a manufacturer have to count the boxes, make sure that the count of boxes matches up with what they expected to receive, and enter the data into their inventory. This costs the workers time, which costs the retailers money. This affects the price that consumers eventually pay for the goods. Any technology that makes that process more efficient makes the entire system more efficient.

Interoperability, like transaction costs, is largely about what happens at the edges of a complex system. Greater interoperability in the system will often make movement among different parts of the system and among multiple systems (such as the computer system of the wholesaler and the computer system of the retailer) cleaner and more straightforward. That reduces transaction costs, cutting out deadweight loss and making the system overall more efficient; in theory, at least, everyone benefits. The retailers' work inventorying the goods is just one step in a long supply chain that includes manufacturing, shipping, and unpacking before the goods make it to consumers. Each part of the chain costs both time and money. Shortening, speeding up, and smoothing out the chain will make everyone better off.

We could classify the systemic improvements that interoperability makes in a multitude of ways: according to which component of a given system they act on, what level of the system they work at, and so forth. But the most important classification is a slightly more abstract one, and one that is much more a spectrum than a binary difference: whether an improvement is more radical or marginal.

A marginal change is one that happens at the edges of the state of things. It is something that can be broken down into a smaller series of gradual steps. Radical changes, on the other hand, involve jumping past the gradual steps into an entirely new framework.

A radical improvement strikes at the core of a system. In the context of interoperability, radical improvements come about when the elements are not interoperable at all or are only barely interoperable. We make them interoperable, and we find ways to harness the interconnection. Consider the effect that bar codes had with regard to retailer-supplier relations: because sales trends were immediately available, retailers and suppliers had an incentive to work together to target consumer interests. With the addition of EDI, the supply chain was revolutionized. Just-in-time manufacturing and inventory management allowed for a great reduction in inventory costs and a better response to consumption trends.

Efficient peak pricing on the smart grid—the energy infrastructure of our near-term future—is another good example of a potential radical improvement that can stem from a higher level of interoperability. Smart meters will be able to give granular information about how much electricity is being used on a minute-by-minute basis. They will be able to pass this valuable information to the electric utility, which will in turn be able to set electricity prices much more accurately than they can today; right now, utilities often have to guess at pricing, based on historical data. This more accurate mode of pricing will make the system more efficient because consumption decisions will be better informed.

Note a common thread between these two examples. Radical changes can often take place where there is a dearth of information. In these cases, technology can either allow better record keeping, or it can make that information more readily available so it becomes usable. With new data, it is possible to optimize systems in new ways.

This phenomenon of radical improvements has a great deal to do with increasing levels of data generation and connection in society at large. As technical layers of systems become more complex and less separable from their data layers, new developments in systems are less likely to happen on purely technical levels and more likely to happen on data levels. As a general rule, so long as the cost of sorting through information is low, more information is better, because it allows all the involved parties to arrive at the solution that is best for the group.

Marginal improvements are less important than radical improvements to the discussion of interoperability and efficiency. But marginal improvements are much more common. Whereas radical improvements add interoperability where there is almost none, marginal improvements change the degree of interoperability existing in a system. For an example of marginal improvements, consider the impact of a chain of coffee stores allowing customers to download an application to their smartphones, preorder their drinks, scan a bar code at the register, and move through the system more quickly. The basic processes of ordering, paying for, and delivering a coffee are the same. The mechanics of electronic payment are little changed. But the system is made to operate more smoothly and effectively for all involved.

Marginal change and radical improvements through interoperability can also come about in tandem. The railroad system offers an example of such a combination. Early on, track width was not standardized across the United States. Train systems could not intersect. Passengers had to dismount and remount; cargo had to be taken off one train and loaded onto another. In order to stitch together a system from the existing rails, it would have been necessary to find a solution that would allow trains to travel on tracks of various widths. In this case, there was already some degree of interoperability present. There were ways for people, goods, and trains to transition from one type of track to another, but it was inconvenient. Eventually, America's railroad network owners decided to standardize track width instead. This decision to render the system interoperable allowed the system's owners to save time and money and expanded the reach, productivity, and speed of trains. Goods, people, and ideas could spread more freely across the breadth of the United States.

This example illustrates how blurry the line can be between marginal and radical changes. It is especially important, when considering whether a given change was marginal or radical, to choose at what level to examine the changes. In this case, from the perspective of train movement, the standardization of track gauges was a radical change; a train that would have had a hard time going from the East Coast to the West Coast all of a sudden could. But on the level of transportation technology—moving people or cargo from coast to coast—it did not make nearly as big a difference. Seen from this angle, it is a marginal, not radical, change.

One specific type of interoperability, which combines the power of compatibility with network effects, is worth special attention in the context of systemic efficiencies. The way this combination generates efficiency is unique. Recall the problem of the typewriters and computer keyboards. Most US computer keyboards are on the QWERTY standard. (The relatively few that are not QWERTY are standardized to use an alternate system, the Dvorak Simplified Keyboard.) Most users approaching a new keyboard know what to expect. This means that they can start typing immediately rather than having to adapt to the new keyboard beforehand.

This standardization on QWERTY obviously generates efficiency among computer users. There is global benefit to this standardization, because users around the world do not have to hunt for keys every time they change computers. Without this standardization, international commerce would move at a slower pace. Although Luddites might be thrilled about that, it is indisputable that efficient use of computers has revolutionized society, and losing that efficiency would make our world a very different place.

Network effects and compatibility, a form of interoperability, combine in this case to reduce switching costs. Systems that involve multiple players working with each other—consider a retailer who might be ordering from one of many manufacturers—can operate unpredictably. In such systems, players do not want to have to incur costs in switching from interacting with one player to another—a particular kind of transaction cost called a switching cost. Participants in such systems will often agree upon a standard in advance so that nobody has to pay switching costs. Usually, adopting that standard has some initial cost but returns money in efficiency gains in the long run.

In the case of QWERTY, it is clear that it is better for the community of computer users if people do not have to readjust their typing habits whenever they change computers. The switching cost in this case would be the time and effort to learn how to use a new type of keyboard. The loss of efficiency would be in the slowdown in computer communications.

The world of retail again provides an excellent example of this type of standardization. Retailers use EDI to encode their orders into a standardized format that manufacturers' computers can decode automatically. Just as bar codes make it easier for retailers to check that manufacturers sent the right goods, EDI makes it much easier for manufacturers to understand what retailers are trying to order. Switching costs are lower for both buyers and sellers.

This gain in efficiency saves a great deal of time in the ordering and shipping stage of the retail supply chain. Retailers can order goods to meet their short-term needs more precisely. As long as they accurately judge what they will sell, they can capture a great deal more utility in the system. And this only works because of network effects. The retail supply chain would work much less well if retailers and manufacturers had not, at some point, agreed on a standard technology to use for EDI, paid to adopt it, and implemented it.

Bar codes and EDI both serve as examples of the collective action problem being more or less solved through the introduction of a certain degree of interoperability. Nearly all manufacturers now use Universal Product Code (UPC) labeling and scanning technologies, and industry has benefited immensely from it. Although businesses still face difficulties in implementing EDI and achieving perfect interoperability, it has been implemented widely enough to generate substantial benefits for almost all participants.

The question then remains, Why hasn't another technology, radio frequency identification (RFID), which could offer similar efficiency gains, taken off in the commercial context the way bar codes and EDI did? There is plenty of evidence to suggest that RFID is a superior technology. Drivers who travel on toll roads are already familiar with RFIDs: they are the technical magic behind E-Z Pass and other systems that let cars fly through tolls without stopping to pick up a ticket or to pay cash. The same technology, when embedded in goods or in the pallets that carry them, can allow for highly efficient tracking of materials. Likewise, one might ask why bar codes have not been entirely replaced by QR codes—those square graphic images that now often appear on advertisements and that can be scanned from a smartphone. QR codes can store vastly more data than bar codes do, at no obvious extra costs (other than switching costs). Why might firms be willing to take a chance on one system (bar codes and EDI) but not another (RFIDs and QR codes)?[1]

There are many potential reasons why new, efficiency-improving forms of interoperability might succeed in one case and not in another. The adoption of new standards typically entails start-up costs for all participants. These are often not too substantial, but in some cases they can be significant. Sometimes adopting a new technology requires so much investment, in time or money, that the possible benefits generated by achieving interoperability fail to outweigh the costs required to achieve it in the first place. This is particularly true in the case of collective action problems, where the benefits are often uncertain. If, as in the case of RFIDs, multiple parties would need to adopt the technology at the same time to make the network effects appear for all parties, it is possible that no one will take the first step forward. Increased risk tends to create a further disincentive to the adoption of the new technologies.

The failure to adopt RFID is much less surprising in light of the history of bar codes and EDI. Both these earlier technologies faced difficulties much like RFID's in the process of their adoption. And in both cases, the new technology was widely adopted only when a large organization chose to invest in implementing it. Those investments allowed them to cross the tipping point so that other firms would find it worthwhile to implement them.

The adoption of bar code technologies was gradual at first. Despite the self-evident value of creating a standardized UPC labeling system, manufacturers were hesitant to spend money to tag merchandise if they were not sure retailers would be scanning them, and retailers were hesitant to install scanning systems if manufacturers might not be tagging their merchandise. Bar codes only truly took off when large retail chains such as Walmart required their suppliers to use UPC labels on merchandise. EDI adoption followed a similar but slower path. Eventually the large retail chains put their muscle behind it, and it too was broadly adopted and implemented.

The story of RFID is similar with respect to the roadblocks involved. Although wide adoption of RFID tagging would allow firms to become more efficient, omitting deadweight loss in manufacturer-retailer relationships because the materials can be tracked so easily, doubts about whether RFID will ever be widely adopted prove self-fulfilling. Thus the system operates less efficiently than it would otherwise. RFID technologies have existed for many decades; the first patents related to RFIDs were issued in the 1970s and 1980s. Early on, venture capitalists and technology firms rushed in to offer RFID solutions to large manufacturers and retailers. The technology was hyped as offering enormous efficiency gains. In 2004, major market actors such as Walmart and government agencies began to insist that suppliers use RFIDs to mark pallets and other shipping equipment. The problem was that early adopters of RFIDs did not initially receive the return on investment that they had expected. Thus the next firms considering adopting RFID technology have less incentive to make the same investment, as they, too, would get a lower return on their investment that the hype around RFIDs might have led them to believe. The vicious cycle is obvious and hard to break.[2]

There may also be a more subtle dynamic at play in the case of RFID. The problem, once again, is lock-in. One successful interoperable technology in a system—in this case, bar codes—can hinder the adoption of a more efficient one. Of course, lock-in seems unintuitive: why would a firm fail to adopt a system that is more efficient than their current one, just because they already have an efficient system in place?

The difference between marginal and radical changes sheds some light on this question. Imagine that there is no sort of merchandise tracking system in place in our market of manufacturers and retailers. The market participants are offered a choice between implementing bar codes or RFID. RFID will cut costs to a greater degree, and so they pay the costs to implement RFID, reaping all the associated benefits. But this is not the world we live in. In our world, bar codes preceded RFID by many years. Now, instead of capturing the radical benefits of implementing an entirely new interoperable system, firms only capture the marginal benefits of implementing RFID over bar codes. RFID generates fewer benefits than bar codes did when they were first introduced, but the process of adoption of RFID requires paying the similar switching costs to make the change. This differential makes it less profitable, on a marginal basis, to implement RFID than it was to implement bar codes in the first place. As such, the likelihood of broad adoption of RFID is lower than the likelihood of bar code adoption before they became widespread.

RFID might eventually succeed. QR codes seem to be gaining traction, especially in the context of mobile applications, the social web, and advertising, at least. Outside intervention could help nudge the adoption of both forward. Other actors with substantial clout, such as the US Department of Defense, have found uses for RFID tagging, forcing at least part of the market to adopt tagging. Other discoveries—for example, a way of printing chips less expensively in general—might also help bring the costs down. As some parties are forced to use the technology and as costs fall, the certainty of benefits rise and the switching costs diminish. This increases the chances that a positive cascade will happen, and RFID tagging will become commonplace.

RFID tagging helps demonstrate one other important principle in the story of interoperability and collective action problems. Generally, with respect to interoperability, the role of the state is to stand back and let the market evolve on its own. However, sometimes a failure to adopt interoperability solutions can be classified as a market failure. In such cases, light-handed government intervention could be valuable and desirable.

This kind of government intervention can help overcome what is at core a collective action problem where nearly everyone stands to benefit when it is solved. Collective action problems often come about because none of the parties with the power to affect the problem has the ability to survive a failure. If RFID tagging fails, the firms involved might be crippled or even become insolvent. That is why the state, which is substantially more durable, comes in. A government experiment in the use of RFIDs might help a market emerge. If relatively mild government intervention can push the collective action problem to a positive equilibrium, rather than a negative one, government action is worth taking.

Interoperability can lead to great gains in the efficiency of complex systems. In many cases, introducing interoperability to a system where before there was none allows revolutionary changes to take place. And, as we can see in the case of order automation and other innovations that emerged later in the retail supply chain, introducing interoperability can have long-term payoffs. Even if the system is already partially interoperable, increasing the degree of interoperability in the system can often have benefits of its own or enable surprising developments in the system.

The sort of efficiency created by interoperability is best viewed through the economic lens. Neoclassical economics predicts gains in economic efficiency when transaction costs fall. There is a tight parallel between this theory and the notion that complex systems can become more efficient through higher degrees of interoperability.

But just as interoperability is not an unalloyed good, efficiency in systems is not necessarily always a good thing. In many systems, it is preferable to have certain levels of stopgap measures and human oversight. These create some deadweight loss by slowing the system down, but they make the system better off overall by averting major problems. Nonetheless, most deadweight loss created by inefficient systems is entirely unnecessary, and in many systems, it is certainly a good thing when interoperability reduces barriers within the system, enabling it to run more smoothly, more quickly, and at lower cost.

The most dangerous aspect of interoperability is that, taken to an extreme, it can create extraordinarily high levels of complexity. These complex systems sometimes grow so immense and intricate that we as humans have a hard time getting our minds around them, much less managing them on an ongoing basis. As we will see in Chapter 8, the effect can be catastrophic for entire economies.

CHAPTER EIGHT

Complexity

I n seeking efficiency, we humans have a tendency to go too far. It is possible for a system to grow to a level of efficiency through interoperability that leads to spectacular failure. Complex systems can become too hard to understand and even harder to manage. The addition of high levels of interconnectedness means that harm can spread like contagion through these complex systems, and we do not have in place the mechanisms to stop it.

The domino effects that led to the global financial crisis of 2008 and the difficulties of responding to it provide an instructive example. Between 2008 and 2011, the fall of legendary financial services firms like Lehman Brothers, the failure of regulatory institutions the world over to prevent major economic losses, and the fear that entire European nations could default on their debts came together in a sustained and perfect storm. The damage continues to reverberate around the globe. No one, except perhaps the most determined advocates of globalization, could help but be concerned about our highly connected economic system. Financial systems worked phenomenally well together in good times, but that interconnectedness also meant that the markets' failure was of a far vaster scale and scope than it would have been otherwise.

Scholars and businesspeople will be studying this economic crisis and systemic collapse into the next century and beyond. History records many cases in which events in one state affected the people of another, from war to famine to economic hardship. But what is different about the story of this collapse is what it demonstrates about the risks associated with the degree and type of deep and extensive systemic interconnection—a deep form of interoperability—among societies and economies around the world.

In certain parts of the financial sector, we have reached a level of systemic complexity so extreme that no single human being can possibly comprehend it, much less manage it effectively. This complexity started at the level of the financial instruments involved. In the case of the 2008 crisis, the key instruments were the novel insurance products built on the shaky foundation of debt almost certain to go bad. These financial instruments enabled a handful of bankers (and some of their clients) to make billions of dollars a year, while putting at risk the livelihoods of billions of ordinary citizens. The complexity ran up through the balance sheets of the biggest financial institutions in the world, from banks to cities to nations of every size. The complexity of the system had grown so great that no one had a prayer of accurately modeling how all the pieces fit together. Despite the wake-up call delivered by plummeting stock markets and failing institutions, we have done little to manage this complexity since the banks began crashing in 2008.

What makes our world so complex, in part, is the vast web of invisible links that connect people and systems among and across cultures. With great effort and anticipation, we as a global society have created layer upon layer of highly interconnected systems. We have found ways to ensure that virtually anything can pass from place to place, despite different kinds of infrastructure and systems of law. We have created mechanisms whereby a man in Shanghai can place an order for a product in London, which can be assembled in Bangalore and arrive a few days later on his doorstep in China.

People, goods, money, and ideas flow more or less freely through nearly two hundred countries around the world.

In most respects, this high degree of interconnectedness brings social benefits; at its core, it makes possible economic growth and cultural exchange at a global, systemic level. It can help drive innovation, competition, and creativity among firms and people within and across individual economies. It can help ensure that consumers have more choices and pay less for what they acquire. As a general matter, this interconnectedness emerges through bottom-up market mechanisms, and occasionally via more forcible, top-down measures such as government fiat.

This interconnectivity, in many ways so desirable, can bring with it drawbacks that were hard to foresee. It was not interconnectedness per se that caused the world's financial systems to shudder collectively and repeatedly between 2008 and 2011; the market is still shuddering as this book goes to print. However, the high degree of connectivity made it possible for the problems of one part of the global system—say, the government of Greece—to affect the fortunes of people and places far from southern Europe. That is not to say that the extent of the interconnectedness among global financial systems has become undesirable in the wake of these crises. But what it does tell us is that we have to think carefully about what the optimal level of interoperability is across systems. We need to determine proactively how to limit the flow of certain negative effects from one system to another in key moments, especially when a system—like the financial system—reaches a level of complexity that no longer allows for effective ad hoc interventions when a problem occurs.

The crisis that began in 2008 put a spotlight on a major design flaw in the global financial system. The system works extremely well in times of growth and optimism, and it enables the spread of funds, people, and innovations from one market to another. But we were, globally, completely unprepared for what would happen in times of contraction and fear. The system had not been designed with firebreaks that could arrest the spread of harm from one part of the system to another in times of crisis. Again, the problem is not with interoperability per se but, rather, with the way the system is designed to deal with the downsides of what interoperability enables. In highly complex systems, some problems can be neither prevented nor managed. But we learned, in all too costly a way, what can happen if a system has far too few levers that decision makers can pull to stop the effects from spreading across many national borders when things go badly wrong.

These kinds of massive failures are not an inevitable consequence of interoperability. Systems can be designed in ways that at least give societies a chance of managing the risks associated with high degrees of interoperability. The starting point is to gain a deeper understanding of the ways interoperability increases systemic complexity.

T he increase in the complexity of systems as a result of growing interoperability tends to follow a rough pattern. The initial state is one of separation. To stick with the global financial example, take as a starting point two nations, distinct from one another, with separate systems of governance and finance. This separation might be purposeful in the first instance. The peoples of the two nations may not wish to interact, or each nation may fear that if it begins to trade, jobs will be lost for its own citizens when the exchange benefits the other nation more than it benefits itself.

Eventually, however, another generation of leaders comes to power and happens to see things differently from those who made the decisions previously. The two nations establish mechanisms of trade, believing that the benefits may be mutual, or at least that both sides will benefit to varying degrees over time. From there, the value of interoperability becomes clearer as people, ideas, products, and money flow from nation to nation and as the financial fortunes of both peoples grow.

At the same time, the interoperation between the two nations gives rise to a new series of problems, unforeseen at the beginning of the shared systems. Whose law should apply, for instance, when a dispute arises between citizens who live in separate countries? Legal institutions need to be established to resolve divorces, broken contracts, and harm caused by faulty products that cross national boundaries. Diseases that existed in one place but not the other now crop up in the second, requiring the introduction of new care facilities, newly skilled doctors and nurses, and different medicines. The financial fortunes that tie the two nations together have unforeseen consequences—for instance, a terrible storm in one country might cause job loss in another. When the number of nations involved is nearly two hundred rather than two, the complexity and risks involved with these harms increase exponentially. The European Union, with its twenty-seven member states, serves as an especially rich case study that gives a preview of the level of sophistication required to manage these complexities.

The same pattern, more or less, occurs on a medium scale in the context of business operation. Most of the examples we draw on in this book have to do with the way information systems work together. These systems are developed primarily by companies like IBM, Google, Microsoft, Yahoo!, Facebook, and their competitors in the global marketplace. The pattern that many firms in the information and communications business follow is similar to that of trading nations.

In the first stage, the businesses traditionally pursue a strategy of exclusion. Firms form, build teams, create a product or service, and work hard to keep others out. The goal is to exploit the value of what one has by excluding competitors. This strategy may still work in some cases and industries, where exclusion of others continues to result in the greatest return for shareholders.

Increasingly, though, businesses are seeing the merits of strategies based on openness. A growing number of businesses are pursuing models that incorporate interoperability as a core principle. More and more firms, especially in the information business, are shedding their proprietary approaches in favor of interoperability at multiple levels. The goal is not to be charitable to competitors or customers, of course, but to maximize returns over time by building an ecosystem with others that holds greater promise than the go-it-alone approach.

The Internet is the simplest example of how this strategy works for many firms and for their customers. The Internet is a network of networks, owned by many and controlled in full by none. On this platform, millions of businesses now thrive. Without a high degree of interoperability at many levels of this complex network, these benefits would go unrealized. Not everyone agrees with interoperability-based strategies, but a growing number of firms are recognizing their value as the power of information networks expands.

But, just as we wish we had built more firebreaks into the global financial system to limit the effect that Greece's failures would have on the rest of Europe and the world, we may come to wish we had designed more safeguards into our increasingly interoperable information and communications systems. It would be much more effective to do so early on rather than after a disaster has struck. Concerns about cybersecurity—worries that the United States is today taking very seriously—make this point most effectively. We have been very successful at linking up and computerizing many functions that are critical to our safety and our economy, including the power grid and transportation security. The notion of electronic warfare, and more commonly electronic terrorism, is no longer a hypothetical scenario. Our systems are indeed vulnerable because they are so interconnected in so many ways.

We do not claim to have all the answers to the problem of how to harness the benefits of interoperability while keeping the risks manageable, especially in the context of complex systems. As with interoperability more broadly, the optimal level of interoperability is highly context specific. And the checks and balances we need to build into complex systems depend on a large number of variables. Massive amounts of brainpower and research resources have gone into the study of complex systems and their manageability over the past few decades. This body of knowledge—a field called complexity theory—provides important insights not only into the limits of what can actually be controlled but also into how complex systems can be designed so that interventions are possible and effective. Many of these so-called leverage points—feedback loops, mechanisms of delay, use buffers, and other techniques—have to do with the regulation of the flow of information within and across systems.[1] A future theory of interoperability, which can also be understood as a step toward a broader theory of information regulation, needs to incorporate these findings and build upon them.

S o far, we have discussed potential drawbacks in the context of complex systems at an abstract, systemic level. But the challenges involved are not issues that only governments and multi-billion-dollar companies have to deal with. The complexity factor also affects us as consumers in our daily lives. Consumers regularly express concern that the high level of interconnectedness created by interoperability can make it very hard to make informed choices and decisions—for instance, when purchasing consumer electronics products or services. The broad range of available systems, applications, and components—many of them connected through invisible links—can lead to *information overload* that the average consumer finds hard to digest. And it is often even more difficult to understand the consequences of our decisions as consumers of technology, in particular.

Take online services as a case in point. Young users, especially, find it very hard to fully understand what happens to their data once they accept the terms of services of, say, a social networking site provider that might systematically share—because of high levels of interop—various pieces of information with third parties. Much of the current debate about the protection of privacy in the Internet age is an expression of worry about the increased levels of interconnectedness among various services, including aggressive marketers, and about the natural limits of what we can cope with as users.

The increased complexity of technological systems may also lead to decreased *reliability*, with negative consequences for consumers. Although this drawback is not an inevitable result of interoperable systems, flaws in complex systems might be difficult to fix quickly. For example, system X, upon which systems Y and Z have come to rely, may have flaws, and solving the problem may need cooperation among X, Y, and Z. If the problem affects consumers, they may find they cannot call upon a single place to get it fixed but, rather, need to call upon more than one.

Consider a home technology user who buys hardware from two companies (her laptop from Hewlett-Packard and her home router from Cisco), software from another (Microsoft), and Internet access from a third (Comcast). If she has a problem with getting Internet access to her home, she might well be bounced around from call center to call center as she seeks to solve the problem, because the service relies upon these systems all working together. Is Cisco on the hook, or is it Comcast's problem to resolve? The consumer is very often left in limbo among these competing parties, none of which want to take responsibility for supporting for these interoperable systems.

As systems that interoperate scale up, the level of complexity may continue to rise. This is one area where an open standards approach to interoperability holds out the most promise. In complex, interoperating systems, we ought to prefer those means whereby problems can be solved by firms and for consumers as seamlessly as possible, without always having to come back to a single firm to fix an underlying problem. Open standards could mitigate, though not outright solve, this issue, insofar as problems might be solved collaboratively, with multiple stakeholders having a right to a seat at the relevant table. For instance, customers should not have to worry about whether their document files will be compatible with the government's document systems, not only on day one but over time. Open standards processes can limit the extent to which any single firm is gaming the system for its own benefit, at the expense of the public as a whole.[2]

Another area where high levels of interop in complex systems may have negative effects for consumers is *accountability*. Much as with the reliability concern, the potential problem here is not inevitable—it is merely possible. And it is related to the extent to which many firms have a hand in providing a service to a given customer. Consider again the example of the customer who bought services from four firms for her home computing setup. Her problem of being bounced around from call center to call center is just the beginning. The concern is, Who is accountable, and responsible financially, if something breaks when many firms are offering a piece of the service to the customer? In a complex, highly interoperable environment, it may not be clear what the relative obligations are of each firm involved in providing the product or service. This is not a grave problem, but it may lead to fights down the road, especially as the seams between the services get blurrier and blurrier. Another way to see this problem is to realize that there is often little distinction between a "product" and a "service" in the web environment. Software, for instance, is increasingly a service one pays for like electricity or legal help, rather than a product that one buys at Staples like a pen or a notebook.

Information overload, the reliability concern, and the potential lack of accountability are three examples of how highly interconnected systems can reach a degree of complexity that affects consumers negatively. From the standpoint of the consumer, too, we need to think hard about how to determine the optimal level of interoperability. System designers, engineers, business leaders, and governments need to work in concert to implement mechanisms that will enable us to deal proactively with the negative effects of a highly interconnected world.

Interoperability can help us reconceptualize the many ways our lives are intricately connected across geography and political borders. AIDS, severe acute respiratory syndrome (SARS), and bird flu illustrate the potential for transmissible diseases to harm more people than ever before possible. The eruption of a volcano in Iceland in April 2010 led to aftereffects in airports and markets around the world as an ash cloud floated across Europe. Fears that a big-spending Gulf state might not be able to pay its debts sent shock waves through global markets. When such events, otherwise more easily contained, take place in an interoperable environment, they can have a destabilizing impact much further afield. The abstract notion of globalization, with its many benefits and costs, becomes clearer, and more immediately present in everyday life, through the lens of interoperability theory.

Interop also helps us understand the complexity of the global economy, how firms relate to one another, and how we as customers fit into that constantly changing picture. The question of who is responsible for what kinds of problems can be solved through careful contracting among the parties and consumers, but it will be increasingly hard to be clear about the boundaries of each party's responsibilities as systems grow more complex and interconnected. For example, one can imagine a scenario in which a user's identity is misused by a third party with whom the user has no contractual relationship. In such a scenario, the rights of the user with respect to the third party might be adversely affected. These concerns show, once again, that the problem is less a consequence of interoperability per se than of its concrete implementation. Good lawyering is at least part of the answer between and among firms. But consumers will still face the problem of knowing to whom they should turn for relief.

Having looked at the many costs and benefits of interoperability, the next step is to understand the options for achieving interoperability and for designing it in a thoughtful way. This book is neither a celebration nor a condemnation of interoperability. We argue in favor of a new series of design principles to help us achieve optimal forms and levels of interoperability in the context of complex systems. Society needs interoperability, but systems must be designed to harness its benefits while minimizing its costs—and without going too far, without creating a system too complex to be managed. The stakes are extremely high.

PART III

Solving for Interop

CHAPTER NINE

Getting to Interop

T his is a good moment to take stock of where we are in our study of interoperability. We have established that interop can be a driver of competition, innovation, consumer choice, and economic growth, among other good outcomes. We have seen that it is an important part of the modern economy, in areas ranging from finance to transportation to information technologies. We have determined that the dynamics of global information flows, combined with the power of network effects in the Internet age, are making interop increasingly important. Despite the potential drawbacks to which it can give rise—including security concerns, risks to privacy, and high levels of complexity—working together to increase interop is generally better than not. We know also that we should aim for an optimal level, not a maximum level, of interop. In this and the chapters that follow, we shift toward exploration of two related questions: How do we determine, in any given situation, what that optimal level of interop is? And how should we go about achieving it?

The first hard problem is to set the interop target. It is important to determine, up front if possible, the optimal level and nature of interoperability in a given circumstance. The key point is to balance the costs and benefits of various levels of interop rather than simply assuming that more interop is inevitably a good thing. There is no single type or amount of interop that makes sense in every instance.

The past tends to complicate this task of setting a target level of interop. Projects to establish interoperability rarely begin with a clean slate on which a leader can divine what "optimal" looks like and then design a process to achieve it. The issues that bedevil this process include path dependency and legacy problems. Initiatives to increase interoperability almost always have to deal with past decisions that shaped a given interop problem and might be hard to undo, as well as with existing technologies and practices that are deeply embedded in a complex system, as we have seen in the context of air traffic control. In most of the cases we have studied in our research, the only feasible approach is to decide what level of interoperability is desirable and to work from an existing baseline. By contrast, the last big examples that we explore here—cloud computing, the Internet of things, and the smart grid—are new enough that it is possible to think about interop more or less from scratch.

Once we have determined the optimal level of interop, the problem grows harder as we turn to how to achieve it. The question of "who" is central to the question of "how." In some instances, it is reasonable to rely on market incentives alone to achieve an optimal level of interoperability. Most of the time, it is a big mistake to place blind faith in the market to establish interop. In most situations, it is necessary for the government to be involved, one way or another, to help move things in the right direction. In all cases, we need to think proactively, as early as we can, about the most promising ways to achieve interop—a process we call interop by design. The instruments that we have to fix a problem after the fact, such as competition law, are blunt, and because they take so long to have an effect, they are often ineffective at accomplishing interop-related goals.

The reasons why it is so hard to get to an optimal level of interop in complex systems are many, but they primarily come down to complexity and the number of people involved. Many different people and institutions have an interest in the state of interop in the examples that we have studied. In the contexts of air traffic control, emergency communications, computer operating systems, and social media, the level of interop is determined by many decentralized and typically uncoordinated decisions. These decisions are made by a heterogeneous group of people, companies, and governments, and they reflect technological preferences, data formats, organizational issues and workflows, and contracts and other legal arrangements, among other concerns. These many actors, making this broad range of decisions, must be coordinated in a strategic way toward the optimal level of interop, which is no mean feat.

Interop is also hard to achieve because it is often a moving target, especially in information-related contexts. A key characteristic of today's technology environment is the speed with which changes occur. Just as a musician or an athlete needs to stay loose in order to perform well, so too must those seeking to establish interoperability remain flexible and open to change as they set policy in this fast-changing context; rigidity can result in disaster. Only a few years ago, we would not have been able to envision social media platforms such as Facebook or Twitter, apps that run on mobile phones, or cloud-based services that let companies outsource entire IT departments. We live in a "quicksilver" technology environment, as the US Supreme Court put it in a recent case, and that fact must inform our decision-making processes.[1] And the most important changes often have nothing to do with technology. Organizations and legal frameworks, for instance, can be in a state of near-constant motion. These shifts put continuous pressure on a given state of interoperability. This pressure makes interop sometimes hard to sustain. The case of libraries, which we will take up in Chapter 12, illustrates this problem of maintaining interop over time.

As we consider how to get to interoperability, we begin with the range of approaches available to accomplish it. There are many paths to interop, each of which comes with certain advantages and disadvantages. Broadly speaking, we can distinguish between private-sector-led and government-led approaches. It is only rarely the case that one single approach—say, for instance, cross-licensing of intellectual property between two private companies to make their systems work together properly—is sufficient to achieve higher interop across complex systems.

Our research shows that most of the time, the best approach involves working with a *combination* of instruments to get to optimal levels of interop. The blended approaches that we advocate typically mean capitalizing on a combination of market forces and state action. Although we believe in the force and efficiency of the market in general, our research demonstrates that governments can and should play a very important role in breaking down barriers and achieving optimal levels of interop in the digital age. A pure laissez-faire approach to interop rarely works out well.

I nterop comes about through a wide range of modes, involving a wide range of actors. There is neither a single approach nor a single configuration of actors that will get us to an optimal level of interoperability in every case. The converse is also true: there are many ways to get to interoperability, many actors who can participate, and many tools in the toolbox that the full range of actors can put to work.

The means by which interop can be achieved can be mapped onto two dimensions. One dimension is defined by who leads and who follows. The leader or prime driver of an interop process might be one or more companies, a nonprofit consortium, or a government agency. But many other actors can participate in interop-related processes, even if they are not in the lead. In the fast-growing social media arena, for instance, these actors include the end users of a given product, whose preferences help guide the decision making of the product's developers. For example, large groups of Facebook's hundreds of millions of customers often make suggestions to the company that lead to increases (or decreases, as in the Beacon case that we considered in the context of user privacy) in interoperability. Consumer interest groups, companies, industry consortia, standards-setting bodies, and governments all can play a role in bringing about interoperability.

The second dimension is the mode by which these actors operate. These modes range from unilateral decision making by a single agent to highly collaborative, distributed processes. The hardest question related to interop is how to choose among these modes and how to decide which actors should play which roles.

No single approach to interop, nor any single configuration of actors, can successfully be applied universally. Each interoperability challenge is case specific, and so it is essential to take a close look into the dynamics of the problem and the market involved. It is imperative to ask detailed questions: What is the exact type of interop issue we are confronting? Which market forces are at play? How mature are the systems and the technology involved? What are the risks and benefits of government intervention? By answering these questions, we will be better able to select the right instruments from the interop toolbox. Each of these tools comes with its own characteristics in terms of effectiveness, efficiency, and flexibility.

Some believe that the private sector should always lead the way to interoperability, to the exclusion of the state. These free-market purists see no role for government in setting interoperability. Their argument is that the state only gets in the way of progress, stifling innovation as it moves into the marketplace to adjust levels of interoperability. Others, mistrusting the private sector, want early and consistent government involvement to ensure that the public welfare is protected as interop-related decisions are rendered in the marketplace. The market, say the interventionists, favors narrow private interests and fails to account for the needs of society as a whole. There is merit in each of these perspectives, but neither one has it entirely right.

In general, we favor *blended* approaches that draw upon the comparative advantages of the private sector and the public sector. A single approach to interoperability rarely succeeds in a complex system. The best approach to optimal levels of interoperability almost always involves a mix of tools used in a range of ways. And this blend of approaches probably changes over time as circumstances change: the state may play a bigger or smaller role at different moments in the history of a given complex system.

T he private sector has led most approaches to interoperability. But there is an important caveat to emphasize here: even where the private sector has led, it did so against the backdrop of rules and standards set by the state. The state always and inevitably sets the context in which private actors are permitted to function. Even if the state does not act directly (say, by passing or amending a law that mandates interoperability), it has been involved indirectly by setting background conditions—legal and otherwise—against which the private-sector players are operating.

Private-sector-led approaches have a series of advantages over state-led initiatives. For instance, the private sector can be quick off the block. No approval is needed and no specific group must be included in order to get going. Private-sector-led approaches lack certain characteristics of state-led processes. For instance, there is typically no classic enforcement power to back up any given approach to interop that the private sector comes up with.

Private-sector-led approaches to interop can vary enormously. Let's start with an example, such as the social web, where the main driver is not a public policy goal (such as health care reform) but, rather, consumer demand and economic incentives. Technology companies have a range of tools available to make their products and services more interoperable.

These private-sector-led approaches fall, roughly speaking, into two categories. First, technology companies can take unilateral actions to increase the interoperability among their own products and services, at least at the technology layer. Second, companies can work together with others in the industry to increase the interoperability among competing components and systems.

In the unilateral approach, systems are designed from the beginning with interoperability in mind. Google Maps, for instance, is a web mapping service application that exemplifies our interop by design approach, in this case undertaken by a single company. Google Maps now underlies many popular location-based services, ranging from route planning to urban business locators for countries around the globe. Not only does Google offer map-based services directly to the public, but it has also invited and enabled third parties to offer services based on their platform as well. By using the Google Maps API, a reasonably skilled computer developer can embed Google Maps into any website. Over 350,000 websites interoperate with

Google Maps; many of them overlay other information on top of it. For example, one can map the location of pizzerias in a city or provide information about crime statistics for different neighborhoods. Google Maps has been such a success that Yahoo! and other companies now provide similar interfaces for their maps. Interoperability by design is one approach by which companies can achieve compatibility among otherwise competing systems.

After unilateral development, standards processes are the second big category of private-sector-led approaches to interop. Standards processes play a particularly important role in getting to interoperability. Standards processes differ from the unilateral approach because they ordinarily require collaboration among a group of private actors, as well as state actors and civil society in some cases. Consider the development of laptop computers. At least 250 interoperability standards are involved in the manufacture of the average laptop produced today. (This number is almost certainly a gross underestimate, because it only includes standards that facilitate technical interoperability. It does not take into account quality, safety, performance, measurement, environmental impact, design, or any other area that might be subject to standards setting.)[2]

Laptop computer standards have had a wide range of origins. The same is true of most complex technology systems. Individual firms can set standards—as in the case of Microsoft's Windows operating system. This type of standard is called a proprietary de facto standard. But standards can also be the result of collaborative efforts. Such so-called formal standards-setting efforts often include international, regional, or national standards organizations. Informal group-led standards-development initiatives, called industry consortia, also play an important role in setting standards. In the case of laptops, the majority of standards were developed primarily by industry-driven consortia (44 percent), followed by those developed by formal standards-setting organizations (36 percent) and by individual companies (20 percent).

Even if state actors are not directly involved, it is a mistake to think that standards processes are entirely controlled by the private sector alone. In the context of laptop computers, the standards processes have taken place very much in the shadow of the law. The law of intellectual property governs how a particular technology developer can establish rights and wield them in the context of the standards-setting process. Competition law governs how laptop makers and software developers interact in bringing new products to market. Consumer protection laws establish a series of limits on how laptop manufacturers and their software partners may offer their products in the marketplace.

Standards processes are often highly contested. A great deal is at stake: companies or persons who set a standard have much to gain from establishing their leadership in the development of a certain system. This gain might take the form of increased market share, enhanced reputation, and the ability to shape the future development of a system. People and firms sometimes fight fiercely, even as they collaborate, in the context of standards-setting processes. Controversies about standards have led to what is known in the literature as standards wars.

The battle of the web browsers is a classic example of a standards war. The fight, in this case, was over which firm would control interoperability of the consumer experience of the web. In 1994, Netscape introduced its very popular web browser and tried to set a de facto standard that would be "free but not free." That is, the Netscape browser was given away for free to end users to encourage broad adoption and thus install a large user base attractive to web masters. However, Netscape planned to earn revenues by charging those who profited commercially from the application of the browser, such as corporations and PC manufacturers.

Although Netscape first appeared to have succeeded in setting this de facto standard, Microsoft changed the game by introducing its Internet Explorer browser. Microsoft's huge advantage as a distribution network— given the reach of its operating system—enabled it to compete immediately and fiercely with Netscape. In the midst of this standards war, a surprising thing happened: the market tipped away from Netscape, in favor of the second mover. These actions also triggered the antitrust investigation of Microsoft in the United States.

Just as battles over de facto standards set by individual companies can lead to disputes, so can collaborative standards-setting processes. Disagreements arise often on matters of substance: which of the different technologies available at a given time should become a standard? Disagreements also arise with respect to process: is the process by which the standard is set fair to all participants?

Take the standards-setting process associated with the Universal Serial Bus (USB). USB is a standard that specifies how peripheral devices such as computer mice, digital cameras, printers, external hard drives, memory sticks, and the like can communicate with a computer or other smart device. A consortium consisting of powerful high-tech companies, including Intel, Compaq, Hewlett-Packard, Lucent, Microsoft, NEC Technologies, and Philips, developed version 2.0 of the USB standard. Observers have claimed that the standard consortium manipulated the USB standard in an anticompetitive way. For instance, other than Intel, no firm that produces chips or microprocessors was included in the process. One of the companies involved is alleged to have used its early access to the specifications of the standard to its advantage. For example, NEC Technologies announced fifteen days before the USB 2.0 interface was released that it had developed the world's first USB 2.0 device.

And finally, the process by which we arrive at standards is itself contested. The merits and demerits—and even the very definition—of certain forms of standards are contested. In particular, *open* standards have been at the core of an almost religious war between proponents and opponents of open versus closed systems.

The core idea behind open standards is that they are developed and maintained via a collaborative, open, and consensus-driven process and made available to the general public. Open standards processes, for instance, have led to a sustainable environment for document formats. In the past, computer users had to choose among programs that did not communicate well with one another, such as Microsoft Word, WordPerfect, and files generated by Sun's OpenOffice products. A series of open standards processes brought together developers of these word processing systems and systems that interacted with them to set standards to ensure interoperability among the range of file formats. The process involves setting standards in the open, rather than behind closed doors, and being open about the intellectual property rights involved. And as standards need to evolve, the process continues in a similar open vein, ensuring that customers and technologists who rely upon the standards do not get left behind as changes inevitably occur.

Proponents argue that open standards maximize knowledge about a given technology and foster competition. Open standards processes can help ensure that innovation thrives over time while reducing the risk that intellectual property rights can get in the way of progress. This open dynamic, in turn, leads to a wider variety of products and services and lower prices. Lower prices will further increase demand by way of what economists call a positive feedback loop.

Open standards processes are desirable from the viewpoint of public policy, but they are not without complication. Skeptics argue that open standards, although often effective, are not the only solution to all interoperability challenges. These skeptics further argue that open standards zealots tend to miss the point: the process may be totally open, but the resulting interoperability may be suboptimal. And not all open standards processes are created equal. There are many approaches, for instance, to the handling of intellectual property rights. A common point of tension is whether open standards should be free of any legal restrictions or whether they can also include licensing terms that are fair, reasonable, and nondiscriminatory.

Despite these objections, it is plain that open standards play a central role in the development of massive communications systems, including the Internet itself. The Internet Engineering Task Force (IETF), for instance, runs ongoing open standards processes to support the development of many aspects of the Internet as it continues to grow. Open standards processes have served the development of the core of the Internet extremely well. The core elements of the Internet have remained largely free of proprietary control in ways that have led to extraordinary levels of innovation and creativity on top of them.

Open standards processes are very often the best approach to achieve interop in the information and communications context. But it is also clear that they are not the only way to do so. Of the market-led mechanisms for achieving interoperability, open standards processes tend to be the most transparent to outside parties. These processes tend to allow for broad participation by a range of different firms and other actors, such as governments. And the interop solutions developed through open standards processes tend not to be encumbered by intellectual property restrictions by any single firm. Open standards processes are certain to play a key role in the development of the complex systems of the future, though as part of a mix of possible solutions.[3]

The state has a role to play in every interoperability process. At the simplest level, there is no such thing as a purely private-sector approach to interop, at least in the information technology business. The state is involved in important ways no matter what, by setting the background rules for how companies must compete and by establishing various limits, such as those established by intellectual property law. The same goes for interop in the context of such complex systems as finance, energy, and health care.

Although approaches like open standards and company-to-company licensing often work well, it would be naive to leave the job of determining the appropriate levels of interoperability solely to market mechanisms operating within the constraints of the state's background rules. The state must sometimes intervene in getting to or maintaining interoperability.

Sometimes the state must be proactive. There are special forces at play in the high-tech markets—such as network effects—that can produce levels of interoperability that are not conducive to competition, at least in the long run. Similarly, market forces do not automatically lead to appropriate standards or to the adoption of the best available technology. And market forces can bring about the hard-to-regulate complex systems that make possible great harm, such as the financial system meltdown.

These complications are reason enough that governments can and should play a role in the process of establishing interoperability. Governments play an important proactive role in fostering interoperability throughout the planning process for a new complex system, such as electronic health records. The state should also be ready to intervene in case things develop in the wrong direction.

A government has several mechanisms at its disposal to increase or inhibit interoperability in the public interest. Some of these measures are proactive, or *ex ante*, approaches, used before a problem arises. For instance, a state can ensure interop by forcing a market actor to disclose interop-related information from the outset. States can also intervene in the market before problems arise in other ways: as buyers (in the case of IT systems) or providers (in the case of health care) of products and services. And lawmakers can establish specific rules that create the incentive for technology companies to work toward interoperable products and services.

The government also has other tools—reactive, or *ex post*, measures, used to correct a situation after a problem has occurred. For instance, a state can bring an action to enforce competition laws, as the governments of both the European Union and the United States have against Microsoft. These kinds of measures are ordinarily slow, expensive, and not all that effective in addressing the problem at hand. They are like the sword of Damocles, poised above the head of the market. A genuine threat of enforcement is necessary to ensure that market players act appropriately in future cases, even if it is an inefficient way to solve hard problems.

Lawmakers can create interoperability by fiat where appropriate. When the importance of the interoperability trumps other concerns, such as the desire for competition or diversity, it makes sense for the state to mandate interoperability, or even to mandate a given standard. These cases are relatively limited: emergency communications, for instance, and rules for transportation are instances in which the public good of having a simple, clear, highly interoperable standard outweighs the costs of using such an approach. The rules for the navigation of boats is one such example, where the state often simply establishes the standards for communications without broad consultation or industry involvement. Air traffic control, as we have seen, is another example. In such cases, however, the state must be willing to allow for periodic review of its rules, to ensure that lock-in does not occur, as has happened in the context of air traffic control. The state needs to be aware also that rules that work at one point in time may not continue to be optimal, especially as conditions change. Rules that are too inflexible can thwart needed improvements to systems like air traffic control.

The state has more specific ways to drive interoperability than might appear at first glance. Its options are not limited to passing new laws or enforcing existing ones. Creativity on the part of the state in helping get to interop can pay substantial dividends for society.

A good and creative way for a government to promote interoperability and increase competition is for it to use its procurement power. Governments are large, important customers of information and communications technologies. The US government, for instance, spends over $75 billion per year on IT. When the government purchases IT equipment, software, or services, its buyers can make it clear that it will only purchase systems or components that are able to interoperate with certain other IT products or services. Many governments have developed coherent interoperability strategies, particularly where it benefits from and uses information and communications technologies to provide e-government services such as online tax filing and applications for certificates. The UK government was an early mover in this regard. The UK government's strategy explicitly calls for interoperability among the systems it procures.

States can also require disclosure of information to competitors as a means of driving interoperability. The European Union's remedy in the Microsoft case, in which it prompted disclosure of technical information to competing firms, exemplifies this approach. When companies create products and services that are not optimally interoperable, governments may require firms to disclose sufficient technical information to enable other companies to build interoperable products. In return, the company with the proprietary technology will receive a fair royalty rate, while other companies benefit from a more level playing field for competition.

The French National Assembly adopted an ex ante version of this type of forced-disclosure approach in the Apple case. Worried about the interoperability problems that undermined access to digital music for consumers, the French amended their Intellectual Property Code. The fear was that technological limitations regarding music access, part of so-called digital rights management (DRM) systems, prevented online users from shifting the digital content they purchased from the online store of one company to the media player of a competitor. Apple's iTunes store, for instance, essentially bars its users from transferring songs purchased from Apple to Microsoft's Zune player. According to the French law, software publishers, manufacturers of technical systems, and service providers can contact a regulatory body known as the Regulatory Authority for Technical Measures in cases in which a DRM supplier refuses to give access to interoperability information. The authority can then order the disclosure of such interoperability information.

The most aggressive, top-down government intervention to increase interoperability in the market is to impose a particular technological standard on companies. Typically, this approach includes not the development of a standard by the government itself but, rather, the mandating of one of the existing standards available in the marketplace. This approach is also the riskiest one for a government to take; it means declaring a "winner" in a marketplace where the government officials may not have the best vantage point to make the call.

NATO, for instance, has taken the especially strong step of choosing a particular standard. NATO made the Open Document Format (ODF) standard mandatory to support interoperability among the various national governments.[4] NATO's approach is sound on one level: its adoption of ODF has promoted the use of a widely open standard for documents and has given a boost to open standards processes. The risk of such direct intervention is that state actors end up favoring a specific standard, which cuts off development of others that may prove better over time. The distinction between using procurement power to demand a certain type of interoperability and selecting a particular technology standard is important.

This riskier strategy, in which the state drives a particular technological standard, also helps demonstrate the variation in regional approaches to interoperability. Common standards for wireless voice and data services, including cell phones, are exemplary in this respect. In Europe and much of Asia, states have mandated specific, harmonized standards for each generation of wireless networks. In the United States, the government has taken a different approach, letting the market decide which standard to adopt. In the context of its e-government framework, the United Kingdom has adopted a series of common standards that already had widespread marketplace acceptance.

The previous approaches to interoperability that we have described are all ex ante; the state can also intervene in interoperability processes ex post, after a certain level has been set. These ex post approaches are often pursued when market forces have failed to provide adequate levels of interoperability on their own. Each of these approaches and tools has costs and benefits, which vary significantly from case to case.

In the category of ex post intervention mechanisms, competition law is one of the bluntest instruments available to the state. Governments use competition law to address certain types of interoperability problems that market forces have created or cannot sort out. Antitrust interventions can operate with considerable effectiveness in the establishment of interoperability in specific fields, such as computer operating systems or the online music landscape.

Competition law is, however, so blunt that it is not a particularly effective instrument in setting interoperability levels. One reason for this ineffectiveness is the simple problem of time. Legal enforcement procedures typically take a long time. Interventions after the fact risk lagging behind market developments. Consider the time involved for the United States and the European Union to bring their massive competition claims against Microsoft based on market dynamics of the 1990s. These disputes ran well into the next decade before they were resolved. Meanwhile, the browser war between Netscape and Internet Explorer, to give one example, took many twists and turns as the legal disputes raged on. The state's claims certainly affected the market dynamics, but they did not "solve" the problems that consumers and competitors faced; the market had moved on by the time the matters were meaningfully resolved. This lag is almost inevitable in cases involving the fast-moving information technology markets.

The existence of this shortcoming does not mean that competition, or antitrust, laws should not be used to bring about interoperability. The most important reason to bring such costly, and often ineffective, actions is their influence on market actors' future behavior. The credible threat of legal action to force a certain type or degree of interoperability can have an important preventive effect. The deterrent effect of large fines or antitrust guidelines can serve a similar function even without large and costly enforcement actions.

Competition law can also help ensure that standards-setting processes achieve results in the public interest. Consider the case of the wireless communications industry. Cell phone manufacturers use different technologies to connect cell phones and their wireless systems, creating a potential interoperability problem. In order to ensure that cell phones from different providers using different technologies can connect with each other, providers and manufacturers of cell phones have engaged in a series of industry-led standards-setting processes.

Two companies involved in this standards-setting process, Broadcom and Qualcomm, compete in the business of producing and licensing technology related to cell phones. Both have developed chips that operate cell phones. Broadcom's chips are based on a technology for which Qualcomm holds a patent. During the standards-setting process, Qualcomm pushed hard for a particular standard. That standard was finally adopted, in part because Qualcomm promised that it would license an essential piece of technology—for which it holds the relevant patent—on fair and reasonable terms. After the standard was adopted, however, competitors claimed that Qualcomm refused to license the technology. In complex litigation, which ended up in a US court of appeals, Broadcom argued that such deceptive actions during a standards-setting process violate the law and that antitrust laws indeed should govern the actions of members of private standards-setting organizations. Broadcom was ultimately successful, and antitrust law did its work, but the process of getting to interop in this case was slow and costly.

The key point is that the state has a wide range of potentially useful options for helping get to interoperability. State actors should think creatively and carefully weigh the costs and benefits of intervention to determine when and how intervention can help guide the market to optimal levels of interoperability in the public interest.

There is no single, best way for private actors and governments to foster competition and innovation by promoting interoperability in fast-changing technology markets. Because of this complexity and heterogeneity, we favor a process-oriented approach to state involvement. Our research shows that each of the approaches at lawmakers' disposal has advantages and disadvantages, depending on the context in which it is applied. We propose a multistep process and a set of guidelines that can help state actors determine the best way to achieve and maintain interoperability in a given technological, market, and social context.

First, the state needs to establish a good reason to intervene. It is entirely possible that the background rules the state has already set have done their job: interop may be close enough to optimal levels without further intervention. The Hippocratic oath applies: first, do no harm.

Second, the state needs to identify and clearly articulate the underlying goals of the legal or regulatory intervention. The objective should not be interoperability per se but, rather, one or more public policy goal to which interoperability can lead. The goals that usually make sense are innovation and competition, but other objectives might include consumer choice, ease of use of a technology or system, diversity, and so forth.

Third, the state needs to consider the facts and variables of the situation. Pertinent factors to consider when developing a public policy–based interop strategy include the maturity of the relevant technology, the structure of and power dynamics on the marketplace, and the timing of the intervention. The state needs to disaggregate types of situations and intervene where the benefits of doing so outweigh the costs. In the most complex technology standards-setting processes, the state is rarely in a position to call a winner among competing technologies. By contrast, where public safety is immediately at stake and the state can set a relatively simple rule to protect people, strong, clear ex ante intervention may make a lot of sense.

Fourth, the state should consider which mechanisms are most likely to lead to a desirable outcome. Criteria should include effectiveness, efficiency, and flexibility. Additional, context-specific evaluation criteria—for instance the risk of possible spillover effects or the time that implementation would take—should be added on a case-by-case basis.

Throughout this process, it is essential for lawmakers to remain open to the possibility that the best approach to interoperability is not the one they started with. More than one approach might be needed, especially in the most complex of situations, such as health care information systems. In many instances, a blended approach may hold the most promise from a public policy perspective. For example, it may make sense to for lawmakers to support a standards-setting process led by private players to help the best, or most open, standard take root. If a state actor later needs to put a thumb on the scale to favor a particular standard, the procurement power can serve this function.

In some instances, the most effective tool for lawmakers is their power to convene stakeholders. The ability to call the appropriate parties to participate in a collaborative, open standards process can be tremendously powerful, even if the government does not make the ultimate decision about the direction in which to lead.

In other instances, the facts may suggest that a single firm can drive innovation by offering others the chance to collaborate—through an open API, for example. If a single firm is successful, it is important to monitor that firm to make sure it doesn't come to abuse its position of power. If a single standard or interoperability process is ascendant, it is important to ensure that lock-in does not set in and stymie further innovation. The monitoring function over time—to keep an eye on the effects of interoperability in complex systems—is an essential role for states in general.

Lawmakers need to keep in view the limits of their own effectiveness when it comes to accomplishing optimal levels of interoperability. Case studies of government intervention, especially where complex information technologies are involved, show that states tend to be ill suited to determine on their own what *specific* technology will be the best option for the future. Firms typically know much more about technological development, and about its costs and benefits, than regulators do. In contrast, government intervention in cases where a powerful technology company deliberately reduces interoperability for anticompetitive purposes—as in the Microsoft example from the 1990s—is a good thing. Enforcement of competition laws, although a crucial threat, cannot be the only solution, and in the end, competition law may function better as a threat than as an enforcement mechanism.

The law can work both in favor of and against interop. In the hands of the most skilled officials, the law can help lead to optimal levels of interop; in other instances, the law is not needed and is best kept on the shelf. The relationship between law and interop, though, is more complicated than that picture alone might lead one to infer. As we will see in Chapter 10, the relationship between law and interop is bidirectional and is extremely important for the way complex systems develop.

CHAPTER TEN

Legal Interop

A s societies become more deeply connected through trade, information, and cultural exchange, the importance of interoperability among the legal systems that regulate the global flow of information continues to grow. Just as we work toward optimal technical interoperability when data flow across systems, we need to be smart about the design of laws when information travels across jurisdictions.

These are early days yet in interoperability among laws. The notion of breaking down legal barriers to enable people and countries to work better together is, in many cases, more aspirational than it is descriptive of reality. It is a hard job for national legislators to craft interoperable legal rules that create a level playing field for the next generation of technologies, economies, and cultural exchange. The future success of emerging complex systems, such as cloud computing, will depend not just on market forces but also on a well-developed legal framework. This legal framework must establish trust and legal certainty for both users and providers of future information systems.

The relationships between interop and the law are many, complex, and tangled. As we have seen, the law can help establish, adjust, or maintain interop. At the same time, interoperability is also a feature of the legal system itself, which we call *legal interoperability*. Legal interoperability, broadly defined, is the process of making legal norms work together across jurisdictions. This interoperability may occur either within the legal system of a single nation-state—consider federal and state legislation—or across national lines. Like technical interoperability, legal interoperability is not a goal in itself but, rather, a means to one or more ends.[1]

The relationship between law and interoperability is multidirectional. Higher levels of interop are often the product of carefully designed legal interventions—or, at least, are fashioned in the shadow of the law. An example in this category is the enforcement of competition law against powerful technology companies trying to leverage their market power by excluding competitors. The forced disclosure of interoperability information as a matter of consumer protection is another. Conversely, interoperability itself can prompt calls for new laws to address its effects; it may also lead to the adjustment or reinterpretation of existing legal norms. As an example, consider the relation between interoperability and privacy: interoperability in the technology world is giving rise to fresh concerns about data privacy for individuals. In turn, European states are considering amending their privacy laws to include a "right to be forgotten" that could span across jurisdictions. The changes wrought by higher levels of interoperability in the technology sector are prompting calls for new forms of legal interoperability.

Looking at these interactions between interop and law, policy makers in our digital and global era should aim for higher levels of legal and policy interoperability, as well as technical interoperability, for three important reasons. At the most basic level, legal interoperability can reduce the costs associated with doing business across borders. The point is simple: it is more expensive, for instance, to develop and deploy a web service designed to comply with many different national data protection laws than it is to create one that conforms to a single global standard. More broadly, recent studies suggest that legal interoperability—particularly in the fields of trade and business law—drives innovation and economic growth by making countries more competitive. Finally, increased levels of legal interoperability among multiple jurisdictions can lead to better laws that foster the development of fundamental values and rights, such as information privacy and freedom of expression.

States have several options for increasing levels of legal interoperability. As in the case of technical interoperability, the point is not to make the systems all the same but, rather, to make them work together in particular ways. It is not necessary for states simply to turn over all legislative authority to the United Nations or to create a raft of new international laws that govern all jurisdictions. There is no chance that every cultural difference can be smoothed out through harmonization of the law. Nor should it be the goal to create one uniform "world law." Jurisdictions compete productively against one another, and learn from each other, through the creation of heterogeneous legal and policy regimes. We need to aim for interoperability among legal systems at an optimal, rather than maximum, level, just as in other interop challenges.

The paths toward higher levels of legal interoperability are many and varied—and can also be rocky. Each situation calls for an in-depth analysis of the various technological, market, and other factors that shape the particular international law and policy issue that is up for consideration. Emerging issues in the information business include how to manage the growth of cloud computing and the smart grid, each of which we will explore in detail later. Policy makers need to craft interoperable legal and policy frameworks as they manage these potentially far-reaching new developments without running roughshod over local norms and values.

L egal interoperability enables the flow of goods, services, and information across political systems as diverse as those of the United States, the Democratic Republic of the Congo, and Japan. Shared legal norms related to food safety, for instance, allow the flow of food products across jurisdictions without requiring every legal system to become the same.

Interoperable requirements regarding the qualification of professionals, such as physicians or lawyers, illustrate the power of interoperability to facilitate the flow of human workforces across nations without requiring every licensure system to become precisely the same.

The challenges of legal interop in the Internet age are ordinarily much more complex than coming to a basic agreement as to how food products, or physicians for that matter, pass across national boundaries. As a result, more often than not, global Internet companies have to contend with noninteroperable legal frameworks. High levels of legal interop in some fields, such as Internet law, are still the exception rather than the rule. The problem is growing harder as more and more data are stored in the cloud, in systems that broadly distribute data across multiple physical locations.

The most famous case in international Internet law exposes some of the difficulties of getting to an optimal level of legal interoperability. The case, which spanned many years and occupied courts in both Europe and the United States, had to do with Yahoo!'s introduction of an online auction system in France. The Yahoo! France case powerfully demonstrates what can happen when companies whose business model relies on the global Internet are confronted with divergent local laws. Though it is very much a web 1.0 story, from the early days of Internet business, the Yahoo! France case points to many of the issues that loom on the horizon as the web and digital technologies evolve.

From its base in California, Yahoo! operated an auction business that sold goods around the world from a single website. Sellers on Yahoo!'s auction site offered thousands of items of Nazi memorabilia for sale. A French antiracism and anti-Semitism organization brought a lawsuit against Yahoo! alleging that the company was violating French law by providing access to these materials through the www.yahoo.com website. The key controversy centered on the fact that the sale of this type of content is banned under French criminal law but is protected as free speech under US constitutional law.

It should come as no surprise that the French court applied French law and agreed with the plaintiffs that Yahoo! was, in fact, acting in violation of the French rules. The French court ordered Yahoo! to deter and prevent access by French residents to auctions of Nazi memorabilia on its site. Yahoo!, in turn, sought free speech protection in US courts. After many years and many trips back and forth, Yahoo! failed to defend its actions in the courts. Today, Yahoo! and other US firms are prohibited from operating auctions that sell materials that violate French law to French citizens.

The outcome of the Yahoo! France case is not optimal from the perspective of trade, perhaps, but it is illustrative of the problem that today's global businesses face. One of the biggest and most persistent challenges of operating in a global economy is the need for a firm based in one jurisdiction (say, the United States) to observe many different local laws. It is rarely possible for a multinational corporation to develop a single product offering, whether on the web or in the form of physical goods, that can satisfy the varying legal requirements everywhere around the globe. The costs of this local variation are enormous.

At the extremes, two options present themselves. First, as a global society, we could seek total harmonization of the law in order to foster trade across borders. This is a mere fantasy, of course; there is no chance that all laws will become the same everywhere. The unlikelihood of harmonization is easiest to see in the context of a culturally sensitive topic such as free expression. There is no chance that First Amendment protections enjoyed in the United States will become the norm any time soon in China or most of the Middle East and North Africa, for instance. At the other extreme lies total fragmentation, with legal systems so different from one to the next that economic and cultural interaction are impossible. This extreme, too, is implausible, given how globally interconnected our personal lives and business systems have become. Legal systems must make it possible for a Japanese woman married to a German man to divorce him. Banks must be able to allow the repayment of loans across jurisdictions. Legal systems that are neither perfectly harmonized nor perfectly fragmented can work.

The golden mean between these two extremes is legal interoperability. The world's legal systems have sophisticated mechanisms for legal interop in some respects and not in others. For instance, legal systems rely upon well-worn mechanisms to determine which court is in charge and which law applies when goods or people travel across the boundaries of nation-states. But these instruments, together with the international agreements aimed at helping resolve conflicts among laws, have proved to be less effective in the digital age than they were in the analog era. For example, most newspaper publishers in the analog world have no trouble complying with local laws where their paper is bought and read in print format. Once that same publisher offers its news online, however, it is much harder to determine how to ensure that the publishing company stays within the law of the nearly two hundred countries around the world where people might access the publication. The Dow Jones publishing empire learned this lesson when an Australian businessman sued them for alleged defamation. Joseph Gutnick contended that an article published in the United States but also offered online on the website for *Barron's* (owned by Dow Jones at the time) harmed his reputation in Australia. An Australian court ruled that it had jurisdiction over matters of this sort, which opened publishers around the world to new forms of potential liability. In the Internet age, interoperability among the laws of different jurisdictions becomes at once more important and more complex.

Lawmakers and policy makers, as well as businesspeople, ought to care a great deal about legal interoperability. The strongest reason: plenty of evidence suggests that legal interoperability, especially in the information economy, drives innovation, competition, trade, and economic growth. For instance, when China entered into the World Trade Organization in 2001, it had to change a great many laws and enact a number of new ones to satisfy the demands of its trading partners. China has made large-scale changes, for instance, in its system of intellectual property law. Though Chinese law is not the same as intellectual property law in the United States or Europe, it is dramatically closer today than it was a few decades ago. A recent Chinese study shows that however onerous these changes might have been at the time, they led to a surge of foreign direct investment in China. Companies that had been unwilling to do business in China previously because they feared their intellectual property rights would be disregarded are today more open to investment in China. Though the systems have not become the same with respect to intellectual property rights, the degree of interoperability has risen between China and its trading partners, such as the United States and many European states.

In addition to promoting trade and economic growth, higher levels of legal interoperability can help secure freedom of expression and foster other fundamental human rights. For instance, the improvement of child labor laws in developing countries is at least in part due to the desire to attract outside investment and contracts from multinational corporations. As human rights activists and companies looking to outsource have applied pressure to governments in Vietnam, China, and India, those countries' protections for children have grown more stringent. The same is true when it comes to free expression. Internet service providers, for instance, have gained protections across Europe in the form of "safe harbors" that limit their liability when they offer Internet services. These safe harbors protect the free speech and privacy rights of Internet users across Europe more clearly and effectively than in the past.

Just as for technical interoperability, legal interoperability is a matter of degrees rather than an all-or-nothing prospect. Higher or lower degrees of legal interoperability may be appropriate in different circumstances. Rules that restrict the creation of and access to child pornography, for instance, should be harmonized, or made as similar as possible, from state to state. This high degree of legal interoperability in child pornography laws can help to deter those who would harm children, which is universally abhorrent, and improve enforcement across national lines. In other instances, however, laws should only be harmonized at the level of basic principles, leaving the details to be fleshed out by national legislators. Such lower levels of interoperability are characteristic of many areas of law within the European Union, where so-called directives are used to create a level playing field among member state laws while still giving national legislators leeway in implementing them. Privacy laws, which bear upon issues that are highly culturally sensitive, are an example of this lower level of legal interoperability.

Even where "law on the books" is roughly harmonized, there is often a great deal of variability in terms of how it is enforced across countries and courts. The net result is a lower level of interoperability as far as "law in action" is concerned. The World Intellectual Property Organization (WIPO) Internet Treaties are a case in point. These treaties were enacted to level the playing field regarding copyright law in the digital age. Though national legislators around the world have signed on to the same treaty, they have implemented the provisions in different ways. One of these two treaties contains a provision prohibiting the circumvention of technological protections for copyrighted material. States vary, for instance, as to whether they see the circumvention itself as the only violation or also consider it a violation to make equipment available to others to circumvention technological protection measures. Courts have also varied in their interpretation of the treaties' provisions. The result is a medium level of legal interoperability.

Though occasionally confusing for those seeking to do business across state lines, a medium level of interop can be good policy. (Lawyers love it: medium levels of legal interoperability can create business for lawyers who know how to interpret the nuances from one jurisdiction to the next.) The benefits of this medium level of legal interoperability, seen in the context of the WIPO Internet Treaties, are meaningful. If the proper balance is struck, states can maintain a strong degree of local decision making over aspects of the law that matter to them, while creating ways for people and businesses to work together across national borders.[2]

There are multiple ways to achieve higher levels of interoperability among legal norms. Each approach comes with strengths and weaknesses that need to be matched to the particular problem at hand. Under the top-down approach, international mechanisms are designed and enacted by large bureaucracies, such as agencies of the United Nations. The International Telecommunications Union (ITU) is one such UN agency. It manages, among other things, the phone system by ensuring that a call from one country can connect to a phone line in another. The ITU handles global technology standardization and radio spectrum allocation. It also plays a role in harmonizing cybercrime laws across the world. The ITU has had a central place in information and communications worldwide for more than a century. Top-down approaches to legal interoperability, however, have not been the norm in the Internet era, as member states have typically feared too much centralized control of communications networks.

Legal interoperability can also come to pass through bottom-up processes. More often than not, legal interoperability is the result of a step-by-step process, through which institutions have worked together over a long period of time to agree on certain principles. The EU Data Protection Directive is an example of legal interoperability resulting from the preparatory work of many institutions. The Organisation for Economic Co-operation and Development (OECD), a highly influential international body based in Paris, has helped bring people together to set privacy norms at a global level. The OECD's privacy guidelines are not binding, but they have served as an expression of common aspirations among member states. The OECD has consistently helped drive interop in privacy law by hosting important conventions that help the relevant communities come together and set the stage for coordinated lawmaking.

For simple matters, such as the allocation of country codes to ensure that telephone calls are sent from one jurisdiction to another, the top-down approach can work well. The level of certainty helps people and businesses connect, and the costs to innovation of a single solution, managed by an international body, are not high. With more complex matters, such as the interoperability of privacy laws across cultures with very different social norms, the bottom-up approach is much better suited. The bottom-up approach often involves coordination, but not harmonization or management, by central bodies. This bottom-up approach, if well implemented, can allow innovation to flourish internationally without squashing local norms.

We have argued in favor of hybrid approaches to legal interoperability, neither pure harmonization nor pure fragmentation. The most revealing example of this middle approach to legal interoperability is the grand experiment of the European Union. The development of the EU is the most extensive test bed of this sort in human history, involving soaring successes and crushing failures.

The development of the European Union helps us understand the goals, modes, benefits, and challenges of legal interoperability. The institutional setup of the European Union is premised heavily on legal interoperability as a means to establish economic integration. In turn, the EU's approach to legal interop can help create solidarity among states and citizens without requiring each state to become subsumed in a common culture.

The Union's single market enables the free movement of goods and factors of production across member states by employing multiple forms of legal interoperability. The mechanisms that result in the highest level of interop are those EU regulations that have general application, are binding, and are directly applicable across all member states. EU directives, as noted earlier, represent a lighter touch in the approach to legal interoperability. Directives mandate the result to be achieved, but the forms and methods of implementation are left to the national authorities, leading to lower— but often extremely effective—levels of interop across the region.

The EU directive demonstrates how best to achieve medium levels of legal interoperability across different countries. Its effectiveness shows that legal norms do not have to be absolutely identical in order to work together—a moderate degree of "harmonization" often suffices. EU directives show that legal interoperability is not binary; like technical interoperability, it falls on a continuum.

The use of a directive also illustrates some of the challenges encountered when aiming for legal interoperability. A measured approach toward legal interoperability that leaves some degrees of flexibility often moves quite slowly or fails outright. The EU Data Protection Directive, as in the case of all directives, had to be "transposed" by member states into internal law after being enacted by the EU. It took three years from its enactment for every member state to implement it—and eighteen years from the time of the initial OECD recommendations in 1980.

Another example—the EU Copyright Directive (EUCD)—illustrates another problem: the directive system can lead to insufficient interop. The EUCD has achieved harmonization regarding the scope of copyright protection. But it failed to create legal interoperability on the aspects of the law that some activists care most about, the parts of the law relating to the limitations and exemptions—such as the "right" to make a private copy—of copyright. Those who advocate for the blind, for instance, note that the level of interop achieved through this directive work well for the copyright owners but work less well for those who rely upon exceptions to the general rule. As a consequence, the EUCD creates an equal playing field for copyright protection, but it does not enable the cross-border sharing of books for the blind that include features like narration and digitized Braille. As the example of the EUCD shows, it is not sufficient to get the level of interop right in one respect but not in another.

The EU is such a rich interop story because it teaches important lessons about the different ways legal interoperability can be achieved, about the many barriers that need to be overcome, and about the failures of interop. The EU example also illustrates the idea that legal interoperability, once established, needs to be managed and maintained. The European Commission, for instance, must closely monitor the process of national implementations and has the power to intervene when EU member states do not play nicely. Legal interop is not a matter of set-and-forget; it requires ongoing diligence and care.

There are downsides to high levels of legal interoperability, just as there are for other types of interop. States sometimes do not wish their legal regimes to interoperate with those of other states. The areas of privacy and free expression are two instructive examples. In the case of privacy law, states often want to limit the flow of information about its citizens to other jurisdictions. The United States, for instance, may want to have increased trade with China but may not wish for that trade to allow data about US citizens to flow into China, which has very different rules for privacy and free expression than the United States does.

Sometimes, friction in the form of low levels of legal interoperability may be desirable from a public policy viewpoint.

In the information technology field, one of the most important considerations for states is cybersecurity. Leaders are extremely concerned about the public security implications of our highly interconnected systems. These concerns include a fear that the information technology systems that we rely upon for so many activities may be disrupted by aggressor states. Imagine, for instance, computer hackers employed by one state unleashing a virus on the unsuspecting computer users of another state at a crucial moment. Russia is accused of cyberattacks on Estonia, China is accused of cyberattacks on the United States, and so forth. Those same state-sponsored hackers might break into intelligence or defense-related systems, find a way to bring down the electrical grid, or divert messages intended for key state personnel in another country.

These cybersecurity risks might appear at first blush to be about the technology and data layers of interoperability, and they are. But they also concern legal interoperability. When a state chooses, as a legal matter, to adopt a particular technology standard, for instance, there may be security consequences down the road. If one state makes plain that it supports a particular document format (as NATO did) or a specific standard related to the electrical grid, the effect of this decision may be to introduce a security risk. That document format or that grid-related standard might have vulnerabilities that hackers might discover by running the same system on their local machines. Also, the fact that the same format is used in multiple jurisdictions might allow a single deed to adversely affect a wider swath.

In 2009, researchers at the University of Toronto uncovered just such a security breach, affecting embassies and other sensitive sites in countries all around the world. Called GhostNet by the savvy researchers who discovered it, the attack began in China—amid suspicions, but no evidence, of government involvement—and reached as far as computers in more than one hundred countries in all, including the Dalai Lama's official facilities. The hackers exploited standard systems that made it possible to launch a vast spy ring from a single location in China.[3] The reach of this hacker attack was made far greater than it could have been in the past because the systems involved were interconnected and because so many people were running similar computer programs across the affected areas. In this sense, legal interoperability to some extent created the conditions that allowed for such a wide-ranging attack.

A second angle to the cybersecurity story is the extent to which some states might aspire, through legal and market mechanisms, to control standards that govern the workings of complex information technology systems. Governments other than that of the United States have privately worried for years about the potential advantage gained by having so many dominant technology players based in Silicon Valley and elsewhere around the United States. Nations might see it as being to their advantage to participate in standards wars and to pass protectionist rules that ensure that computing systems are based on home-grown standards that might be less vulnerable to exploitation by unfriendly governments. The surveillance of cell phone or landline connections, for instance, might be easier for countries like Canada because RIM (the makers of BlackBerries) and Nortel Networks (formerly a leading a developer of computing equipment used around the world) have been based there. The same might be said of the United States, because so many smartphone and computing systems providers are based there. States without such technological dominance in the private sector might therefore prefer lower levels of legal and technical interoperability.

The raging debates over cybersecurity help illuminate some of the dark sides of high degrees of legal (as well as technological) interoperability. As our world becomes more and more interconnected across national borders, the ability to introduce threats from afar is growing, as cyberattacks make plain. The choices we make with respect to legal interoperability can have public security consequences in a highly interconnected global economy.

On balance, we expect a trend toward higher degrees of legal interoperability, especially with respect to information and communications technologies. One driver of this trend is likely to be cloud computing.

As our lives become mediated increasingly by digital technologies, we are creating more and more data that need to be stored and processed. The trend for data storage and processing is toward cloud computing, which means that these growing amounts of data are managed somewhere other than where the activities take place on the ground.

We do not have to look to the future to see this trend leading to problems. Disputes over the cross-border flow of data already reveal the growing importance of legal interoperability. The government of British Columbia sought to outsource the administration and management of British Columbians' personal health care information to a US-owned company. The British Columbia Government Employees Union opposed this outsourcing plan and launched its Right to Privacy campaign, arguing that the outsourced data could be subject to the highly controversial USA PATRIOT Act. They argued that the US government could potentially apply the statute to gain access to confidential records about British Columbians. In response to the protests, the government of British Columbia introduced amendments to the Freedom of Information Protection of Privacy Act, which now includes wide prohibition against data disclosures outside Canada. British Columbia's move to block data disclosures outside Canada shows how some states have taken action to thwart high levels of legal interoperability. Canadians are right to erect legal barriers to avoid subjecting their citizens' data to the overreaching provisions of the USA PATRIOT Act.

The case of Yahoo! in Belgium illustrates the same problem from a slightly different angle. A Belgian prosecutor asked Yahoo! (based in California) to turn over user data about e-mail accounts the Belgian government believed were being used for online fraud. Because Yahoo! had no physical presence or data storage in Belgium, the company asked the Belgian authorities to make the request through the appropriate legal channels, using the procedures of a so-called Mutual Legal Assistance Treaty between the United States and Belgium. The government refused. A court of first instance in Belgium sided with the government, imposing a large criminal fine against Yahoo! for not complying with the prosecutor's disclosure request. After a series of appeals, the high court ultimately upheld the lower court ruling. The result is that Yahoo! is subject to Belgian jurisdiction and must hand over data to Belgian law enforcement solely because the website Yahoo.com is accessible in Belgium.

In both the case of health data in Canada and the case of Yahoo!'s data in Belgium, the result of the conflicts was not to change the laws of either of the states involved but, rather, to find a way to resolve the dispute without making the local laws the same. In each of these two cases, the norms of the country where the consumers lived trumped the law governing the company that was interacting with its citizens. Intuitively, this is right: the people of Canada or of Belgium should not be subjected to the law of another country merely because they are interacting with a company that happens to be based elsewhere (in these two cases, the United States). The level of legal interoperability was very low in each of these cases, which serves immediate local interests. The cost of this approach, though, is potentially to frustrate international trade. The result might be that in the future, the British Columbian government might not outsource to a US provider and Yahoo! might not serve Belgians with its e-mail service.

This is far from the end of the story: the idea of the local law winning in every case will become harder and harder to sustain over time. Legal interop problems like these will only become more widespread as we move into the borderless world of cloud computing. Finding appropriate answers to these conflicts will become increasingly urgent. It would be convenient to suggest that all data can be confined to a single location and then merely apply local law. The design principles of cloud computing, however, point in another direction: the wide distribution of data across multiple jurisdictions. These principles enable cloud computing to offer the promise of substantial gains in terms of efficiency and cost savings.

The costs of noninteroperable laws in the highly networked world of global information and communication technologies will increase. Low levels of legal interoperability are likely to have negative long-term effects on innovation, trade, and economic growth. National policy makers and lawmakers should generally aim for higher levels of technical and related layers of interoperability, while recognizing that these higher levels may sometimes come with certain drawbacks. Legal interoperability, if implemented at optimal levels, can allow for the creation of a level playing field for tech companies and users alike, while preserving flexibility and appropriate degrees of diversity.

CHAPTER ELEVEN

Interop by Design: The Case of Health Care IT

ealth care in the United States costs too much money. The United States spends more on health care than any other nation in the world—and yet the premature mortality rate in the United States is much higher than in other developed countries.[1] The high cost of health care and the relatively low quality is due, in part, to massive inefficiencies in the way patient information is shared and managed. Even in relatively simple matters, a very large number of decentralized physicians, labs, clinics, hospitals, insurance companies, medical researchers, and other members of the health industry touch information about a patient. In the United States, the systems that join these many players and the information they need to share are often not interoperable.

Pretty much everyone agrees: the mass digitization of health records in an optimally interoperable system would be a major step in addressing this problem. Such a system could help ensure that accurate information about a person's health is available at the right time, can be accessed by the authorized health professionals and institutions, and is shared securely. President Obama has called for the United States to have a system of electronic health records in place by 2014.[2] President George W. Bush, before him, made a similar pledge. But pledges by heads of state are hardly enough to bring about a solution to this interoperability problem on their own. British prime minister Tony Blair had tried the same thing in 2002 and committed no less than £12 billion to the task, only to fail to meet the initiative's ambitious goals.[3]

Given the high stakes, the billions of tax dollars already spent, and the bipartisan agreement, it is hard to fathom why we have not yet solved this problem. The high cost of information technologies are not alone responsible for the high cost of health care, but it is clear that a better information system could help a great deal. Why has the development of a comprehensive, reliable system of interoperable e-health records become the Holy Grail for policy makers focused on health care reform? Why have the strategies we have deployed to date not worked? Can't someone in the government just mandate interoperable, electronic medical records and be done with it?

There are many reasons why we as yet have no interoperable system of electronic record keeping in the health care industry in the United States. The size and complexity of the system is a primary factor. Amid this complexity, the optimum level of interop in health care information is very hard to define, both because of rapid technological progress and because of the significant potential dangers of interoperability, particularly in the arena of privacy. In addition to these substantive problems, the underlying decision-making processes are highly politicized. Thousands of hospitals around the country have legacy systems in place, which makes it very hard to implement any comprehensive interop strategy. The upfront costs of new technology and human retraining will be massive. In addition, the powerful companies that sell these legacy systems, and therefore have a stake in preserving the status quo, create additional barriers to change. And even if we solve these political and market issues, the creation of an interoperable, reliable e-health records system is a very challenging problem from an engineering, organizational, and legal perspective. These many barriers form a daunting thicket.

In our research for this book, we have explored dozens of complex systems in depth, ranging from transportation to finance to the emergence of the Internet itself, but none is more complex than the health care information case. Each case has taught us that interop often poses enormous design challenges and should not be taken for granted. No case study involves more players, more money, and more problems than the case of information technologies in today's health care debate. For this reason, we explore the problem of health care information and interoperability in greater detail than we do any other single case study in this book.

No single actor can make an optimally interoperable e-health records system come about. It would be ineffective for the government to pass a law mandating a particular standard in health records; the problem is too complex for that. And the market, left to its own devices, is not producing anything remotely like what is needed to meet the challenge. A blended approach, in which industry and government work together, is the only sustainable solution.

H ealth care in the United States costs more than $1.7 trillion per year. That is more than twice the per capita average of the amounts spent by all countries studied by the Organisation for Economic Co-operation and Development (OECD).[4] At the same time, patients get less good care than they should. As members of the Robert Wood Johnson Foundation's Commission to Build a Healthier America put it: "For the first time in our history, the United States is raising a generation of children who may live sicker, shorter lives than their parents."[5] Babies born in the United States are more likely to die before reaching their first birthdays than babies born in Canada, the Czech Republic, Greece, and more than twenty other countries. Everyone agrees that the system does not work as well as it should, especially given how much money Americans spend on it. The scale of the health care problem is staggering.

One reason for the crisis of the health care system, among many, is inefficiency in the way information is handled. The authors of a widely cited RAND Corporation study from 2005 argue that the US health care industry is the world's largest and most inefficient information enterprise.[6] E-health records can be a big part of the solution to this problem of inefficiency. The development of an optimally interoperable system of e-health records would bring with it a series of obvious benefits. The costs that can be stripped from the health care system, for starters, are enormous. The lower costs of providing care—compared to the cost of care provided in a world of paper-based records—can mean that more people get more care that improves their health. Doctors will have improved access to the right information at the right time; patients who need emergency care can be treated properly if doctors can quickly access their full records, regardless of whether their primary care providers are nearby or not.

The idea of introducing electronic health records (EHRs) is one of very few aspects of the health care reform process that has the support of politicians from both major parties in the United States. And it has the full support of the executive branch: not only has President Obama mandated a system of EHRs by 2014, but he has also thrown his full support behind the idea of the government playing an *active* role in making electronic medical records a reality. Obama has pushed the Congress and the Department of Health and Human Services to allocate billions of dollars in funding and human support for the transition to electronic medical records. He has been right to do so, but his administration will have to do more in order to meet the 2014 mandate.

The idea that President Obama has championed, like President Bush and Prime Minister Blair before him, is simple and profound: a better health information system can help us wring cost out of the health care system while improving the quality of care people get. The US government has estimated that Medicare can save approximately $23 billion per year and that private parties can save about $31 billion per year if a broad system of EHRs can be put in place.[7] Two prominent studies have suggested that health information technology systems could possibly save $80 billion per year in 2005 dollars in a system that costs $2 trillion per year.[8] Though some dispute the magnitude of the potential savings,[9] no one argues with the assertion that an electronic records system would be cheaper and more efficient in the medium to long term than our current system, which still relies heavily on paper records. The bottom-line cost savings alone support the transition.

In addition to cost savings, an improved health information technology infrastructure should bring with it enhancements to the quality of care. Patients should be able to get better care in a world of interoperable EHRs, which would provide easier and more efficient access to health information for citizens and for health care providers. The quality of the care that patients get should improve because doctors are less prone to make errors when they have access to the highest-quality information about their patients. Doctors pressed for time—say, in an emergency room—would benefit from having the most up-to-date information and a better presentation of medical history, that is, one arranged by importance rather than by date. The benefits of EHRs are, of course, best realized if the systems are interoperable and widespread—for example, an individual patient will not benefit from an EHR system if her doctor uses it but her local hospital does not. The massive health care provider Kaiser Permanente has put billions of dollars into advanced information technology systems first of all to improve care throughout its system. The Kaiser Permanente system allows doctors to rely upon more consistent clinical protocols and reminder systems that lead to healthier patients.[10]

There are other potential advantages to the widespread implementation of an EHR system. By cataloging the medical information of hundreds of millions of patients across EHR systems, an integrated approach would facilitate the distribution of medical knowledge and research. Interoperable

EHRs could provide access to unprecedented amounts of clinical data that could accelerate epidemiological research as well as the dissemination of new and effective medical practices. An advanced electronic records system could also enable generativity, as competitors developed improved systems on top of the initial innovation.[11]

It is just as important to acknowledge the potential drawbacks to a highly interoperable EHR system. Despite broad political consensus for EHRs, many who support the idea in principle also have deep reservations about the specifics of implementing an EHR system. Some concerns are purely pragmatic: the up-front costs of making the transition will be substantial, the providers of current systems may feel threatened by the prospect of large-scale change, and some doctors fear, or just do not want to use, a new system.

There are other, more complicated reasons that some are wary of an EHR system. Some doctors and hospitals fear sharing too much information across institutions for competitive or legal reasons. Insurers worry that greater transparency into how doctors reach decisions in the course of providing care will expose them (and in turn the insurers themselves) to more liability. Lawyers for hospitals are concerned that EHRs can lead to more disagreements about health care decisions, either because patients may not have sufficient context in which to evaluate the full body of information about their care or because doctors may be more likely to disagree with a colleague when presented with more detailed information about the patient's care. These potential drawbacks to interoperability may persuade some participants to lobby against an approach that most others agree is the way to go forward.

Security and privacy are also among the main concerns about implementing a highly interoperable system of EHRs. One fear, in its simplest (and highly unlikely) form, is that all medical records about all Americans might be housed in a single records center, whether physical or virtual. Such a facility would be a natural target for hackers who want to get access to these enormously valuable data. There is security in distribution: with paper records at every doctor's office in a large country, a thief would never be able to gain access to all the records at once, even though the security of each record itself might be lower than the security of all the records at a central facility. The future of medical records will fall somewhere between these two poles. Electronic medical records will probably reside in many different areas of the network, behind various different forms of technical security protections.

The primary worry that consumers have about an e-health records system is the potential for their privacy to be violated. Although most individuals are comfortable with—even enthusiastic about—the idea that an emergency-room surgeon could instantly access relevant medical facts from their various doctors, many people also harbor profound concerns that the same information could get into the hands of potential or current employers, insurers, and so on. Some details are plainly more sensitive than others: the possibility of information about mental health, sexually transmitted diseases, genetics, and gender leaking into the wrong hands can be especially worrisome. For instance, an interoperable EHR system needs to be able to share information about a prescription with a pharmacist, but the pharmacist should not be able to learn sensitive details associated with the diagnosis. Any interoperable EHR system needs to be designed to take the threat of privacy violations seriously. An interoperable system must be built with inherent limitations on the flow of data about patients. Interoperable EHR systems will lead to the need to update confidentiality and privacy rules for health information.

The move toward an interoperable system of electronic medical records reveals the tension that can arise among competing public policy interests. In this case, the drive toward a more efficient, effective health care system comes up against the long-standing effort to establish strong privacy protections for individual health information. One of the few areas where formal, specific privacy protections exist in the United States is for our health-related information, in the form of the Health Information Portability and Accountability Act (HIPAA).[12] The government's push to expand the use of EHRs places pressure on the rules about health information. The very premise of expanding EHRs depends to some extent on interoperability and sharing of information across contexts. But the information can only be shared in certain approved situations. It is entirely possible to design a system that will honor these privacy-related distinctions, but it will be much easier and potentially more lasting to do so at the outset.[13]

Another drawback of a highly interoperable system of EHRs is the risk of lock-in. Recall that the high level of interoperability of the air traffic control system eventually led to lock-in. To avoid this problem, the EHR system needs to be designed in such a way that it operates reliably and safely in the first instance, while also allowing innovation and competition to flourish. This flexibility has never been more important, especially because changes in information technology are occurring so rapidly. For instance, we are at the very beginning of a health care revolution made possible by advances in the field of genetics. An EHR system that becomes locked in early in these developments could cut off life-saving innovations. The homogeneity risk is enormous in the context of e-health records. If a single approach is mandated too early in the process, or if an initial system is locked in through rules or through market forces, the risk of harm to innovation and improvements in care is high.

On the whole, though, these risks and drawbacks are far outweighed by the many benefits of establishing an interoperable system of EHRs if we get its design right. The puzzle, then, is to figure out what a health information system that embraces the many benefits of interop but avoids the pitfalls ought to look like, and what the best way to make it happen is.

B efore turning to the solutions, it is important to analyze why a comprehensive and interoperable system of electronic health information has not yet come about in most countries. A big part of the reason is the complexity and enormity of the task. A health information system needs to be accessible to many different types of actors and needs to be able to work across many contexts. It requires coordination at all four layers of our interop model: technology, data, human, and institutional. There are barriers at each of these levels.

A further complicating factor is the fact that we cannot build a system of interoperable electronic health records from scratch, which actually makes things harder across each of the four layers. There is already a degree of interoperability in patient health care information. Systems used in hospitals, even when paper based, permit certain flows of information from one environment to another. A patient who is switching health care providers can get the file transferred via traditional communications channels, in hard copy. Researchers are able to combine paper-based information to perform epidemiological studies. The patterns of human interaction are more or less established. The challenge is to get from this limited, primarily analog level of interoperability to a higher, primarily digital level of interoperability across many institutions. And there is an important design question to be answered along the way: how high a level of interoperability is in fact desirable?

One of the existing barriers to interoperability that must be overcome is the engineering challenge of making systems work together so that health information can be shared among them in the right ways at the right times, and not at others. Some systems are hard to make interoperable as a matter of systems engineering (the interdisciplinary method of making complex systems work effectively over time). A vast number and many types of existing files and processes need to be brought together. There can be no interoperability of complex information systems, in the health care system or otherwise, without clever engineers who act with a great deal of foresight.

These engineering challenges are a close cousin to the problem of legacy systems. In most cases where we want interoperability, as in the case of health care, we are not working from a blank slate. Someone has already built information technology systems that, for instance, run the day-to-day operations of a hospital. The prospect of transitioning everyone from a thousand different systems to one single system is potentially overwhelming. And even where a few systems already work together, it might be very hard to achieve higher levels of interop across many more such systems. Interoperability itself can play a role as a barrier to further interoperability, in a curious way. One problem with legacy systems is that lock-in can ensue where previous-generation systems are highly interoperable, and even where a suboptimal degree of interoperability exists. Sometimes it is hard to break and start again. This is why the starting point is so crucial.

The largest barrier to interoperability in EHRs to date, though, is a particular form of competition in the marketplace. Competition, in general, is a good thing. It is desirable to have a level playing field where everyone has a roughly equal shot at coming up with an innovative approach to solving a hard problem. At the same time, competition that is grounded in a winner-take-all philosophy can work against the goal of interoperability. Many firms have been working on EHRs for decades. They have established, through their innovations, strong market positions. If the firms competing in the EHR space see a large pot of gold at the end of the rainbow for whoever comes up with the single standard around which everyone else's solution must be built, then they have little incentive to cooperate with one another.

Finally, the question of who will pay has not yet been answered in full. New information technology systems cost a lot of money to acquire, implement, and maintain. Not all players in the systems will be equally able to make the up-front investment in the EHR platforms and training to use them. The hospitals and other health care providers would prefer the government to pay for the system. Because it is a government mandate, they say, the government should cover the costs of the transition. The benefits of investing in the system will not be enjoyed exactly by those who put the systems in place. For instance, an interoperable system of EHRs may cost physicians more than insurers and pharmacists. But on a proportional basis, insurers and pharmacists may in turn benefit more than doctors from the existence of the systems. And the expensive maintenance of the software and hardware over time will inevitably cost some more than others—and probably not in direct proportion to the extent to which each party benefits from those expenditures.

The future approach to an interoperable system of EHRs must be grounded in an understanding of the barriers that have stymied us to date. The systemic design process needs to address each of these concerns, either through market mechanisms, government mandates, or a combination of the two. A set of incentives needs to be put in place to ensure that a range of actors do their parts to bring about an interoperable system of EHRs, or it will remain out of reach.

T he market, on its own, has not led to a sufficient level of interoperability in the health care information market. To the contrary, the forces of competition are partly responsible for the failure to achieve a system of interoperable EHRs. Despite all the attention on health care reform, there has not been sufficient push to bring the necessary participants together to work toward anything close to an interoperable system. There have been many high-level government-driven initiatives to develop a nationally coordinated system of EHRs.

The failure of these efforts to date demonstrates how hard it is to balance the benefits and drawbacks of interoperability in practice and among many different players with varying interests and incentives. More fundamentally, though, the failure of the market to produce an adequately interoperable health information ecosystem shows that there is a need for a clearer mandate from the US government to get the process more on track. As in other cases we have analyzed, the government's role is not to choose a specific technology but, rather, to ensure that data are able to flow between and among systems and that people are able to work better together across institutions, in the interests of the public at large.

It is one thing to call for stronger government involvement. But it is a different story to get it at the right level and to a desirable degree. If we seek to achieve interoperability in e-health records by pursuing a completely government-led top-down approach, we run the risk of adversely affecting the development of new business models. On the other side of the argument, if too much power is left with the private sector, we run the risk that the victor in a winner-takes-all competitive process will likewise stifle competition and innovation. If the system that emerges is built, for instance, on locking in customers, whether patients, doctors, or institutions, we will all lose. The *manner* in which interoperability is achieved in the e-health records context will matter a great deal to the effectiveness of the system that results.

We propose that the government—acting through the Office of the National Coordinator for Health Information Technology at the US Department of Health and Human Services—should lead a ramped-up design process to put in place a truly interoperable health information system for the United States. A good design for this highly interoperable system must address, up front, the many potential concerns and must be flexible enough to change to address issues that arise after implementation. At the outset of the process, designers will need to consider the type and extent of the interop. Designers need to work out, to a reasonably detailed level, what limits are to be placed on how the interop works in practice. The system must also be implemented in such a way that it is stable enough to last but also flexible enough to take advantage of inevitable changes in technology.

The Obama administration has the right intentions, but the effort to achieve a functioning EHR system will need to be firmer and more directive if we are to meet the president's 2014 deadline. The administration should continue efforts to broker a system by which firms and the state work together to design and create an effective, interoperable system of EHRs. This coordinating role, though, must be accompanied by a stronger stick: there must be an accompanying requirement that certain data can be shared across systems, no matter what. If the market will not or cannot provide this level of interoperability, the government should mandate it.

Central to this design is the nature of the interoperability that the government should mandate, if need be. The basic concept is that systems need to interoperate just as the ATM network does. Systems designers need to ensure that certain transactions will be possible no matter which hospital a patient goes to, much as an ATM machine accepts cards from any bank. One way to think about this form of interoperability is the concept of substitutability. For certain tasks, such as withdrawing funds, any ATM card should be able to substitute for any other bank's card for those same tasks. The same should be true for certain basic elements of health-related data.[14]

No government agency, no matter how powerful, can succeed on its own in meeting the challenge of EHR interoperability, however. Without a great deal of cooperation from the private sector, any comprehensive effort toward interoperable EHRs will fail. Firms in the health information business, for their part, need to take a longer view when it comes to interoperability and openness in a field as big as EHRs. No single firm should, or will be able to, conquer such a complex marketplace in its entirety. Firms should also resist the urge to hold out for as much licensing revenue as possible from preexisting intellectual property at the outset of a process to develop a large-scale system of EHRs. The long view would instead entail working in good faith with competitors and with regulators to establish a marketplace that will be enormous over time and in which many parties can compete effectively. The net effect could be a technological ecosystem that functions in much the same way as does the highly dynamic social web ecosystem, which is today leading to radical innovation—and astronomical profits—online.

This cooperation between the government and the companies involved in developing EHRs is happening to large degree, but not sufficiently to get to a full EHR system in the United States by 2014. The Department of Health and Human Services has set aside $27 billion to help doctors and hospitals transition their systems from paper to electronic medical records. Clinicians can receive as much as $44,000 from Medicare and $63,000 from Medicaid under the transitional program. Despite this taxpayer-funded largesse, some participants have claimed that the government's regulations on the use of the money are too restrictive. The Obama administration has responded to these criticisms by loosening the rules for doctors and hospitals to access the transitional funds.[15] Looser restrictions on the degree of interoperability of the system, though, may increase the difficulty in meeting the 2014 deadline.

It is not enough to say that there is a public-private partnership underway to create a system of EHRs. As they work together, the government and the health care industry still need to address a number of hard issues along the way. For starters, the design needs to incorporate an understanding of how deep the interop should run. As a starting point, the interop in EHRs is not deep at all. Data sharing is often limited in early trials to a few specific types of information—such as test results, inpatient data, and medication lists—and does not include a comprehensive set of data about a patient.[16] Efforts that aim for full standardization of all medical records will fail because of the complexity of the health care system at large and the heterogeneity of the processes involved.

A key part of the government's role should be to determine the optimal level of interoperability and then to drive toward it. The optimal level of interoperability lies somewhere between where we stand today (a very low level of interoperability) and total interoperability (or complete standardization of all records). Privacy and security considerations will push toward less deep, more precise types of interoperability. Concerns about functionality and efficiency of the system will pull in the opposite direction, in favor of deeper levels of interoperability throughout the health care information system. It is at this layer in the interop model that the government's interventions make the most sense: to ensure that certain data can flow throughout the system, limited by appropriate privacy-related safeguards.

The type of interoperability the government should push for ought to look like a standardized indexing system; it should allow for substitutability of different types of information—as in the ATM card example—but not complete standardization. A relative few crucial fields about a patient or a condition would be set and standardized, but other, superfluous, aspects of the records would not be. The system would need to allow easy access to this core information across institutions in such a way that a health care provider could use a networked device, such as a mobile computer or laptop, to easily access the key information to help the patient. For relatively simple matters, physicians could access just what they need to make diagnoses. For more complex matters, they might need to obtain higher degrees of interoperability or access to information in order to make more challenging diagnoses.

The system's design must also ensure preservation of information over time. The health care industry must prepare for long-term preservation and storage of patient records. The industry must come to a consensus on how long to store EHRs, what methods to use to ensure the future accessibility and compatibility of archived data with yet-to-be-developed retrieval systems, and how to ensure the physical and virtual security of the archives. Interoperability can be a driver for the preservation of information, as in the case of document formats (such as Word or Adobe PDF) and libraries. Interoperability should be seen as a contributor to long-term information preservation rather than as a reason not to develop interoperable information systems.

A health information system will also need to be designed to handle cross-border issues that will inevitably arise. As populations become more integrated across political and geographic lines, our information systems need to be designed to work together across these borders. The interoperability of EHRs must expand past technological cooperation to include legal interop as well. Different countries may have divergent legal requirements for content or usage of EHRs, which in turn can require radical changes in how data is managed and stored. Though not a pressing issue for many patients and doctors on day one, these cross-border issues need to be considered during the development phase of a comprehensive health information system.

I n order for an interoperable EHR system to come into existence, the government and industry need to work together on a better system of incentives. Key participants will need to be motivated to implement, use, and maintain such systems.

The basic incentives currently provided by the government are a big help: cash on the barrel to get things set up. Also, institutions may find that having sophisticated health care information systems will attract patients. But further incentives are needed to ensure that institutions implement sufficiently interoperable systems and then put clean, accurate data into the databases.

The government should explore means to encourage health care institutions, such as hospitals, to enter data, in digital form, into a system that allows others to interoperate with the data. There is much to be gained by such a system. The benefits would accrue not just to those who create the system but also to others, such as the insurance companies and pharmacists—a notion that economists refer to as positive externalities. These externalities do not directly benefit the institution (or the doctor) entering the data or maintaining the systems, but they do benefit society as a whole (not to mention individual patients, the insurance companies, and others). In such a situation, the state may need to provide additional incentives for institutions beyond what they would get directly from participating. In addition to subsidizing the purchase of these systems at the front end, the government might mandate payment of a standardized fee when one institution benefits from using data provided by another. Such a fee would give institutions a chance to profit from behavior that helps their patients. A system of this sort might also lead insurance companies and other system participants to pay their fair share of the costs of implementing and maintaining the system.[17]

Likewise, the government and the health care industry need to ensure that innovation can continue. No system developed now should be locked in, blocking the possibility of implementing a better system in a decade. The government should support a mechanism that would ensure that experts review the relevant standards periodically, with a twin mandate: ensure stability and compliance over time, but also ensure that innovation can be introduced where appropriate. The same data must be able to flow through the system, but the approach to making that happen needs to be able to change, and this design principle has to be established firmly at the outset.

The US government needs to intervene in the interop process for EHRs if health care reform is to be accomplished on time and as promised. A central aspect of this particular case study is the role the government itself plays as a stakeholder in the electronic medical records environment. The state, in this case, is both a funder of the system and a direct provider of health care services, through, for instance, the medical facilities of the US Department of Veterans Affairs and through the Medicare system. As a result, the state has unique leverage to intervene in the process of adopting EHRs. As a consumer, the government can refuse to pay for systems that do not accomplish the level of interoperability it has set or that do not enable the kind of data sharing needed.

M ost of the debate around interoperability in the e-health records context centers on issues at the technology and data layers—for instance, the difficulty of choosing the right standards and of solving very hard engineering problems. Interop at these layers is indeed crucial. It can be constructive for the government to help set a standard for electronic medical records and to pay for adoption of particular technologies. But these steps are not sufficient to make the system at large optimally interoperable.

Just as important is the work that must be done to ensure interoperability at the human and the institutional layers. To make the system work in a manner that will help a country like the United States reap the benefits of EHRs, the people who implement and use them every day need to find ways of working that are as interoperable as the technology and data. The relevant institutional parties need to be able to engage, too. Insurance companies, hospitals, the government, and patients all need to be able to collaborate—if not always in harmony, with incentives perfectly aligned, then at least in a manner that enables us all to enjoy the systemwide benefits of lower cost and better care.

The experiences of our European neighbors are, once again, instructive. As a general matter, Europeans are more inclined toward state-driven, comprehensive approaches to solving thorny problems like EHR implementation than is the United States, where leaders have tended to pursue market-driven approaches. Despite numerous experiments seeking to resolve the legal and ethical issues associated with e-health initiatives, as a whole the EU has encountered difficulty in implementing a coordinated EHR effort across the entire Union. Within some countries, such as Denmark and the United Kingdom, EHR adoption has reached relatively high rates—nearly 80–90 percent in some regions of the United Kingdom and 90 percent or better across all Denmark health care providers.

There is no obvious pattern to what the catalyzing factor or factors (for instance, public funding or policy mandates) are that have led to high rates of adoption. For instance, in the United Kingdom, the state publicly funded most implementations of EHRs, though a comprehensive EHR system still eludes them.[18] In contrast, EHRs have been widely adopted in Denmark in a short period of time without firm adoption mandates or extensive public funding. The approach that the Danes have taken has been evolutionary, rather than revolutionary.[19] Denmark, though, has a few obvious advantages: one is its small size, with a population of roughly 5 million. The other is that it has an actual health care system, whereas the United States has never had one.

Our conclusion from the study of these European cases is that a number of external factors beyond policy and funding initiatives affect the adoption rates—and that the cultural and institutional issues matter enormously. In Denmark, for instance, the type of culture within the health care system has affected adoption rates. Trust in the government and in health care providers is thought to be higher in Denmark than in the United States. The reason the push to adopt EHRs has worked is, in part, the intangible goodwill and collaboration throughout a coordinated health care system— and a notable lack of resistance to change. The health care community at large coalesced around a Digital Health strategy, coordinated by a government Ministry of Health and Prevention.[20] These human, cultural factors may ultimately be more important than any technological decision or government mandate.

Health care reform is an example of an area in which we urgently need interop by design. The process of getting to a system of EHRs will require a strong government push if we are to meet our goals without introducing new problems. We need a sound design for an interoperable information infrastructure, championed by the state, that is then implemented, consistently, throughout the wildly complex health care system. The state must ensure that this system is put in place in a way that ensures high levels of security and privacy, rewards innovators and investors, and is flexible enough to grow and change.

A new, interoperable health information system, if we get it right, offers the potential for better care, greater efficiency, and a bright future for innovation in health care. The right level of interoperability will not freeze innovation but, instead, will enable a diversity of approaches to flourish, in the public interest. Making it happen in a system as complex and uncoordinated as the health care industry in the United States will take a vast number of people working together—human interoperability.

CHAPTER TWELVE

Interop over Time: Preservation of Knowledge

Imagine you are cleaning out your apartment and come across a box full of books from your college years. You flip through a few of them and smile at the memory of classes you loved. You remember, too, those classes you tended to skip, for whatever reason. You can tell from the state of a paperback edition of Franz Kafka's *Metamorphosis* that you only ever made it through half the book, and that's being generous to yourself. You remember the famous first line about Gregor Samsa turning into a really big bug, but not much more. Your notes in the margins conspicuously stop partway through the text. The pages in the back half of the book, clumped together as though they just came off the press, remain totally untouched. Elsewhere in the box, a printed transcript reveals that you got a

B-minus for your modest efforts in the course for which the book was assigned. Just deserts, you think.

Packed alongside the books is a pile of old computer disks, held together with a rubber band. They're a little dusty but otherwise appear to be in good condition. Your stash contains both 3½-inch and 5¼-inch disks, carefully labeled according to topic: résumés and transcripts, term papers organized by course, financial records, and so forth from those same college years. You flip them over, rubbing the plastic in your fingers. But there's not much more you can do with them. There's no way to know what you read or what you wrote from that period. The transcript on the 3½-inch disk is locked away; you can no longer access it or share it with a prospective employer or graduate school. Unless you take these discs to someone and pay for conversion, the data on the disks are useless to you. The books and the printed transcript, by contrast, are as useful to you as ever.

It is one thing to create interoperability at one moment in time; it is quite another to maintain it. When it comes to digital media, interop over time is a problem we need to start solving now—indeed, should have started solving years ago, when digital technologies began changing the way we write and read and store information. Information that is accessible today may well be lost forever if we do not establish reliable technologies capable of moving digital information forward. A system that will ensure interoperability over time is crucial to the preservation of our cultural and scientific heritage. Interop over time will also prove important to the proper ongoing functioning of complex systems, like EHRs or the smart power grid.

Libraries face many questions about their role in the digital age. Some people say that libraries are not needed in the age of Google. This attitude could not be more wrong, but it is a common view that librarians have to address. One of the biggest challenges libraries face also represents an important opportunity to demonstrate the crucial role libraries ought to play in the digital era: finding a way to achieve interoperability in digital material formats over time.

To preserve knowledge for posterity, libraries have traditionally collected books, periodicals (such as scholarly journals), and images in printed formats. Today, though, a growing amount of library spending is devoted to acquiring materials in a digital format. In some cases, these materials are so-called born-digital materials—materials that started out digital and are printed or otherwise made analog just as a matter of convenience. Think of a database of information about the stock market or about executive compensation in major US corporations, for instance. The library pays for access to these digital materials, usually for a limited time, on behalf of its patrons. The library, in this example, never gets a physical copy of what it is paying for. Instead of buying a copy, the library pays for its patrons to be able to use the material the publisher is selling. Ordinarily, the grant of rights lasts only as long as the library continues to pay the publisher. When the payments stop, so too does the access.

In addition to purely born-digital materials, libraries also purchase many materials in digital formats—sometimes in conjunction with a printed version. In some cases, a library patron can choose to walk into the library stacks, remove a physical copy from the shelf, and read the book or article from a bound volume as she did in years past. Or she might choose instead to log on to a website from her home, verify her identity, and download the file to read on her iPad.

Libraries are having a very hard time keeping up with this growing multiplicity of formats. Virtually every library finds it too expensive to purchase both the hard-copy and electronic forms of all the materials patrons want. Space comes at a premium, so librarians have to make hard choices about what to keep on the shelves on-site. The decision may well be different for different kinds of materials: in the case of scholarly journals, for instance, most users prefer the digital format, so it is less essential to keep hard-copy formats on-site. Given that digital lending of books is a very new technology, it is too early to tell which format readers will prefer.

Although library patrons continue to use material in multiple formats, digital materials make up an ever-greater proportion of library offerings with each passing year. After all, it is much more efficient as a means of distributing knowledge, particularly for journals. The digital form of journals has already transformed scholarship by offering a much faster way to publish research than in the traditional, paper-based mode. Scholarly databases of journals also offer extremely fast and efficient ways to link research materials to one another, which improves the connections scholars can find. Full-length books may be a different story, at least for the time being; many people still prefer to read them in hard copy, despite a major uptick in digital book sales that began in 2010.

Digitization of reading materials comes with benefits for readers and libraries. For the reader, digital journals and books provide greater ease of access, from any time and any place, and offer the ability to search across sources. For libraries, the costs of purchase and storage, at least in theory, should be lower. The politicians who pay the bills for public libraries, as well as the deans and presidents who pay for academic libraries, see dollar signs: they would like to pay less for the libraries of the future as they move money around to meet other budget demands.

But there are major challenges associated with the digitization of knowledge and the future of libraries. First off, digitization will transform—indeed has already begun to transform—the business model of the publishers and scholarly societies that select, edit, and provide access to knowledge. The perception that digital journals and books will save a great deal of money is largely wrong; the production and storage costs saved are likely to be modest, and there are potentially high costs associated with digital storage and reformatting. The movement toward open access publishing—that is, publishing directly on the web by academics—provides a special challenge to scholarly publishers. And the threat of piracy looms over the publishing industry, just as it has over the music and movie industries. Digital files are more easily copied and shared than analog copies. Put another way, publishers fear that the possibility of massive piracy is greater for e-books than for physical books.

Publishers have experimented with strategies to meet the demand for digital materials in libraries, but have had mixed success. HarperCollins, for instance, tried a strategy that would enable libraries to loan copies of the e-books it published, but no more than twenty-six times (roughly, on the theory that a traditional book would have "worn out" after having been loaned that many times). Their offering was not a success; the idea that a book should "wear out" after it was loaned twenty-six times was widely ridiculed by members of the library community. Librarians and the publishers of journals and books will need to collaborate to come up with a forward-looking business model for journals and books.

The business model question aside, there is another, more pressing challenge posed by the inexorable march toward the digitization of knowledge: how do we preserve digital materials over time? Printed materials have the advantage of being highly durable. Sure, bound books and journals are susceptible to fire. And there have been times in history when physical materials were not good long-term bets, as attested by the fate of the Library of Alexandria or, more recently, of books produced during the period of the "slow fire" phenomenon.[1] For more than one hundred years spanning parts of the nineteenth and twentieth centuries, many books were printed on acidic paper, which began to disintegrate within decades. Nitrate film is subject to similar issues: it proved to be highly combustible, not exactly a property one wants to introduce into a library environment. But as a general matter, the bound book and journal have endured beautifully—for centuries, if not millennia, when made well.

Even as reader preferences tilt toward digital materials, there is no single, default format in which these digital materials are produced. The trend is toward a proliferation, not a consolidation, of digital formats. Given the rate of innovation in information technologies, it is impossible to guess at the new data formats we will favor even a decade from now. Will they be mobile, augmented reality, immersive, viewed by the human eye, or some other form of sensory perception? And will it be possible to preserve the entire look and feel of a digital presentation of materials? It's anyone's guess. But the answer does matter, potentially a great deal, to libraries and to how they function.

Fast-forward a decade or so: Virtually no one publishes journals in the traditional printed form, and books continue to be produced in both digital and analog formats. Library patrons can get all the information they need through increasingly attractive, highly functional digital formats, accessible at any time from any place—and they are very happy with what they get, how it is presented to them, and how they can use it. Devices like the iPad have become the primary means by which most materials are accessed and enjoyed by library patrons. There has been a shakeout in the publishing world, but good writing and high-quality scholarship still find their way into publication. Libraries thrive as centers of knowledge management and as vibrant public spaces where patrons access information and work together in small groups. Librarians prove to be more important than ever for their ability to guide patrons through increasingly vast and complex information environments; the library profession, like the libraries themselves as physical spaces, is strong and vibrant in its new incarnation.

But let us fast-forward another twenty years. None of the digital information that has been acquired by libraries is still accessible. The iPad is no longer on the market. It has been replaced by a completely different type of reading device, sold by a company that did not exist in 2012. The vast majority of digital information to which libraries provide access needs to be reformatted if it is to be enjoyed in this new world. Who pays for the reformatting? Who decides what to reformat and what to leave in the old format? This scenario is hardly unrealistic. Those who bought cassette tapes of music in the 1980s had to buy them again as CDs or as digital files to be able to play them in their cars or on their iPods. We are certain to have large costs to pay in order to reformat today's digital materials so they are usable in tomorrow's devices.

A simple response is to sidestep the problem. One approach is to render information in an analog format, in addition to a digital format, to ensure that it can always be re-created. Put another way, the publisher—or the library—could print out one copy of a text and set it in a vault. Call this the belt-and-suspenders approach. Better yet, perhaps ten libraries agree to house print copies of each journal and each book. Now imagine that a giant magnet swings over the earth and zaps everything in digital format—it's all erased. No problem! We could just go back to the hard-copy archives buried in the vaults and start again. Sure, it would be labor-intensive. But it is entirely possible. And the fact that it would take so much effort would ensure that we only re-create information for which there is a demand, rather than re-creating all information, regardless of its value and provenance. (Yes, there is a risk buried here, too: decision makers might not get the selection process right, and valuable information would be left in the vault. But that is another debate. Another risk is that some materials, such as games or choose-your-own-adventure-style materials may be nearly impossible to "print out" at all.)

The idea of printing out everything we publish in digital journals and books is a backstop, nothing more. It is a badly inefficient strategy. The cost of re-creating materials from the analog backups every time the preferred format of digital access changes would be astronomical—and, in the language of economics, an outrageous deadweight loss. In reality, we probably would not have the will to go back and re-create all this knowledge; some of it, even some valuable material, would be lost forever. We absolutely should print out what we publish and put it into a salt mine to avoid the worst-case scenario. But that is only an insurance policy, not a practical solution to solve the problem of quickly shifting data formats.

The problem of preserving knowledge in a digital age runs deeper still. Even if we can solve the problem of shifting formats, society still faces the problem of ensuring that the data are kept somewhere in the first place. No library, no archive, is successfully looking out for many of the e-mails sent today by presently unrecognized historic figures. Very little formal archiving of the World Wide Web is going on, even despite the extraordinary efforts and importance of initiatives like Internet Archive. The brainchild of Internet entrepreneur Brewster Kahle, Internet Archive and the Wayback Machine are amazing resources, through which one can view versions of the web at certain dates in the past. But despite its name, the Internet Archive is not intended to function as a formal, complete archive of digital materials in the literal sense. There are great histories that will not be able to be written because they will not have access to original, born-digital correspondence between people alive today. It is ironic that cave paintings and papyrus scrolls may allow historical interpretation better than an e-mail sent in 2011 from one major figure to another does.

This is one of the great paradoxes of the digital age. On the one hand, we have too little privacy. Too much about our lives is recorded and shared with others, accessible via search engines and across time. On the other hand, we are saving too little of what we really want to preserve. Again, there is a collective action problem to blame. We have not worked together sufficiently to address this concern, to establish systems for preserving the parts of our present and our recent past that we ought to preserve. Instead, we have delegated to for-profit companies the job of deciding what to keep about each of us, how long to keep it, and what to do with it. We make information of all sorts available in digital formats, and when we do we need to ensure that we are protecting the innocent. Consider, for example, court records. They need careful redaction before they are presented in an open format online and preserved as such for posterity.

The rate of change is partly to blame for this paradox; we are hurtling into a digitally mediated way of living without thinking through all the consequences. And the size of our digital footprint is also partly to blame: surely, we do not need or want to save for posterity every chat message, tweet, blog post, and Facebook status update that every teenager sees fit to post. But we do want to preserve the scholarly record and the primary sources necessary to write the great histories of tomorrow.

It is both important and urgent that we, as a society, find a way to ensure that the information—the knowledge—embedded in each of these digital journals and books and e-mails can be easily recorded, retrieved, and updated over time as information formats evolve. Think of this process as interoperability over time. We need a mechanism that ensures that material created by a writer (or a film producer or a concert pianist) in 2010 can be re-created by the computing equipment of 2025 or 2125 or 2500. The theory of interoperability can help us make digital formats as enduring as print formats for information, for knowledge, and for artistic expression.

As societies, we ought to devote ourselves to establishing a new digital depository library to solve the problem of interoperability of knowledge over time. This idea is a digital-era variant of the old depository library model, which approached the preservation of knowledge by requiring publishers of books to deposit a copy of every book in one or more libraries. This requirement is sometimes expressed as a condition of the grant of copyright, which is a state-conferred monopoly on the exploitation of a creative work for a limited period of time. The main purpose of a digital deposit would be to ensure the availability of these works over the very long term. This approach would be controversial and not without problems, but it represents a structural way to address the problem of interoperability over time.

Here is how it would work. Instead of just contributing a physical copy of every book to a few libraries, all publishers would agree to contribute a digital copy of every book to one or, better yet, to a series of depository libraries. The digital copy would have to be provided in one of a series of standard formats, agreed upon in advance. These formats would be subject to a set of rules, also established in advance, that would require backward compatibility to be worked into their design. In other words, the technologies would be created from the outset with the view that they would be easy to translate into the new formats that would inevitably come into being later on. The idea would be similar to the notion of data portability that makes it easy to export information from a spreadsheet into a common format (such as a comma-separated value [CSV] file, one of the options offered when a user saves a document in a spreadsheet program like Microsoft Excel), regardless of what software is used to produce it.

These data, deposited by publishers (perhaps as part of the copyright process), could be held in many different ways. The best approach would be for one or more libraries to hold the data in trust in a cloud-based environment. The data for each book or journal would be preserved in an authenticated version with an appropriate "watermark" or date stamp and would be kept in this cloud-based environment according to an agreed-upon naming convention so that the work could be found and accessed and so that bibliographic information could be added or extracted. The work would be registered with an independent body, such as each country's copyright office, and would have metadata associated with it. Metadata, which helps explain what the work is about and where it fits in the context of the world of knowledge, is essential to anyone finding the work in the vast digital ocean of information. A hard copy would be printed out and would also be kept at the same depository library or libraries.

This cloud-based approach is not a necessary component of the digital depository idea; files could easily be stored locally at one or more depository libraries. But the cloud-based approach would solve a number of issues. First, it would allow low-cost, long-term storage of the information in several places on the network, which would reduce the likelihood of catastrophic loss if all or much of the data were to be destroyed by an attack or a virus. Second, publishers, though likely to resist the idea at first, could eventually benefit from some of the same properties of the cloud. They would not need to keep as much inventory in warehouses or to pay as much for the physical handling of materials. Third, libraries (other than the depository libraries, which would automatically have a copy under this system) would have the opportunity to purchase works in a range of formats. An option for print-on-demand would enable libraries also to keep one or more hard copies if they wished to do so.

This initial stage—the publication and deposit of the works into a cloud-based infrastructure—is just the first part of the solution, but it will be hard to accomplish. There are high hurdles to be cleared between where we stand today and this cloud-based solution. A broad range of participants would have to agree to collaborate. Libraries and the reading public at large are certain to be beneficiaries of such a process. Many publishers, authors, and their agents would be skeptical of such a plan.

The objections to this plan would be several. A primary concern of publishers, authors, and agents would be the possibility of intellectual property theft. Under this scenario, a copy of all digital books and journals would be kept in the cloud, an aggregation of data that would make a tempting target for hackers and pirates. The music industry, after all, has suffered from mass piracy of copyrighted works in digital format.

But the risk of piracy would not be as high as one might fear at the outset. First, we entrust our financial, health, and national defense information to similar cloud-based systems today; computer security is imperfect, but we already rely on it for virtually all our sensitive tasks. Second, digital files of most books already exist and are offered for sale online through Amazon's Kindle service, for Barnes & Noble's Nook, and so forth. This proposal would not add to the risk in this respect. Third, piracy has not yet proven to be a major concern in many industries, including digital book publishing. Piracy is culturally driven. Although the most popular hit musical recordings have been consistently stolen and shared illegally, classical music and jazz have not been similarly affected. The book-buying community seems unlikely to engage in massive online piracy. Fourth, there are indications that some works can actually benefit from being exposed in digital versions online.[2] Taken together, these facts suggest that the added risk of piracy from the digital depository model is outweighed by the long-term benefits that the preservation of knowledge offers society.

Even if a broad range of participants were to come together to develop a digital depository system, the group would still need to agree upon a process for managing its development and maintenance. The first process to be determined would be how to approve the formats; this would take the form of a standards process—ideally, an open standards process, as we have seen in other contexts. Once the process had been established, specific issues would need to be resolved: which formats are approved for use in the system, what naming convention should be used for storing the copies in the cloud, which method of authentication should be used, and what metadata should be associated with the registered works. Additional concerns will center on special cases that occur more than one might think: preserving formulas and calculations embedded in spreadsheets; pagination and sections encoded in journals in order to facilitate citation and navigation; and appropriate rates of performance to experience time-based media. Other complex issues to be taken up would include international harmonization of the rules and the interaction among the rules and copyright and security concerns. The biggest topics to be addressed are likely to be financial: who pays for the storage, preservation, and reformatting of the authenticated copy.

P ublishing information in interoperable formats in the first instance is necessary but not sufficient to ensure the long-term preservation of knowledge in a digital age. The greater challenge is to preserve this interoperability over time. As a society, we will continue to innovate in the development of data formats. Data storage formats progressed from the 5¼-inch floppy disk in 1985 to the USB thumb drive in 2005; we will surely go through a similar evolution between 2015 and 2030. The hardware to render the data, likewise, changed from desktop machines with floppy-disk readers to netbooks or smartphones with no input mechanisms other than flash memory sticks or web browsers. How do we ensure that the book or journal rendered according to our 2015-era processes can be made accessible in 2030?

The responsibility for and cost of these changes must be shared among those with a stake in the outcome. Publishers, libraries (public and academic alike), and the state all have a role to play. The participants in this process also necessarily include the developers of the technologies involved—including data formats, software, and hardware—as well as cooperative bodies that promote interoperable standards. An example of such a collaborative effort is the Open Archival Information Systems (OAIS) model, established to bring a wide range of stakeholders together to agree on digital archival standards.[3]

In designing this process to achieve interoperability over time, we need to bear in mind a wide range of concerns, most of which we have seen in other interop-related processes. We need to ensure that any system promotes innovation to the greatest extent possible, rather than hindering it. We need to emphasize consumer choice, so that readers are not compelled to purchase equipment from a single technology provider in order to participate in the consumption of information. We need to ensure that the system is sustainable over time. In particular, the system will need to outlive the inevitable failure of individual players in the system, most likely publishers and technology providers but possibly also including publicly funded libraries and states (one can safely assume that there will always be readers, despite the many jeremiads arguing otherwise).

One of the key design features of any approved new data format should be openness to data updated from a previous generation. The obligation to make this updating process simple and inexpensive should fall, by default, on the technology providers. The actual updating of the data should fall to the publishers who wish to sell their works in these formats. The state and the library system should be responsible for keeping up the cloud-based repository of these materials and for storing the physical copies for redundant and archival purposes. If a publisher fails, the state should have the job of using the materials that have been deposited in the cloud (or in the local or physical archive) to re-create the information, if library professionals, and perhaps other subject specialists, deem the information worthy of safeguarding for posterity.

Here is what this process would mean for a book published in, say, 2015. The book, issued by Publisher X in the United States, would be submitted upon publication to the digital depository managed by a central authority. This authority could be the Library of Congress and a series of partners, or it could be an altogether new entity, as long as its expected lifespan is very long—a great university, for example. The book, as submitted to the depository, would be rendered in one of a series of formats that had been approved by participants in an open standards process. Fast-forward twenty years, to 2035. If Publisher X is still in business and still wishes to sell the book, it has an incentive to update the format (and it will contribute the reformatted work to the depository library or libraries). But if Publisher X has gone out of business, the depository library could then take on the job of updating the book from its original, 2015 format to one of the approved 2035 formats, if the book is worth preserving. The depository library would most probably bear the responsibility and the cost to reformat it, but the state might also pay reformatting costs, or a mix of libraries could pool resources to ensure consistent reformatting of such works. Either way, for materials published through formal mechanisms, such as books and journals, a reliable system would be in place to address the problem of interop over time.

A primary objection to this proposal of a digital depository is that it may not be necessary: the market may take care of this interoperability problem on its own. After all, publishers have the incentive to make sure that their works are accessible over time; they survive by selling them to a public that needs to be able to read them. And information technology companies stand to make a great deal of money by being able to sell the gear involved. It is possible that the free marketeers are right. But it is highly unlikely.

The best evidence that the market will not work on its own to solve this problem is that it is not working right now. We are well into the digital revolution in publishing, and information stored in digital formats is currently being lost every day: some materials, such as e-mail and web-based works, are not being stored in the first place, and other materials are trapped in old formats that cannot easily be accessed. The market is already failing us in the important task of preserving knowledge in a digital era. There are promising small-scale efforts at digital preservation, but these initiatives do not have widespread support from the necessary players. As time passes, the risk of proceeding without such an interoperable system in place is growing.

Market-based approaches to interoperability of digital materials over time are vulnerable for many reasons. One is that no one has the proper incentive to spend the money now to solve a problem that causes harm at some time well into the future. A second is that for-profit companies are often less stable than libraries and governments. Publishers tend to be for-profit concerns. Over the course of history, they can be expected to change hands, to consolidate, to change business models, or to fail outright. If libraries move to a model of leasing information in digital format from for-profit entities—on an "access" model only, without owning any of the actual texts—one has reason to wonder what will happen if one of these companies fails. It is too facile to presume that another firm will simply buy its assets. The possibility exists, at least, that the firm failed because its information could not be sold profitably, and yet the information may nevertheless be worth preserving for scholarly, historical, or cultural purposes over time. The cloud-based depository library proposal is one way in which libraries, universities, and states can protect themselves—and the public at large—from the market's probable failure to make good decisions about what information is worth preserving and what can be relegated to the dustbin of history.

A final problem with the free-market approach is that companies will often agree on what needs to be done but refuse to cooperate to do it. Set aside the problem of interop over time; we have a problem of interop today when it comes to e-books. Apple and Amazon both sell e-readers. Both companies agree that a standard format for e-books would be a good thing. But at least to date, they have not been able to come to terms on a way to cooperate. Those books you downloaded to your Kindle are no good in Apple's reading application, and vice versa. You need to download Amazon's Kindle application onto your iPad in order to read your Kindle books on the iPad; there is no way to read the books you buy from Apple's store on your Kindle. This problem is hard enough to solve on day one. Over time, it is only going to get trickier, not less difficult, to solve.[4]

For this range of reasons, private actors cannot be counted upon to handle the preservation of knowledge over time. The stakes are too high. The state must play a role in this area. That said, as in other interoperability challenges, the state should not run this process in its entirety. It should help convene the relevant market participants to establish the system that will provide for interoperability over time. Then it should turn its attention to preserving copies of published materials through an updated version of traditional national libraries. Many states, such as South Korea, have developed a national digital library of one sort or another; the United States has not—yet.

Librarians ought to play the central role in this process, a role that happens to be closely related to the core function of libraries. Librarians are charged with ensuring that we record, categorize, store, render accessible, and preserve information, both immediately and over the very long term. Librarians are well aware of these problems; what is missing is a lack of collective action across all groups of interested parties. And some of the best librarians work on this problem. If anyone is up to the task, it is people like David Ferriero, the archivist of the United States, appointed by President Obama to preserve the nation's historical materials. Under Ferriero's leadership, the National Archives and Records Administration is focused on the issue of digitization and preservation of the heritage of the United States.

Most major libraries are taking steps on their own to preserve digital information, but these efforts are not systematic or widespread enough to address the issue over the long term. In addition, such efforts tend to be field-specific initiatives. Within the field of legal information, for instance, an alliance called the Legal Information Preservation Alliance (LIPA) has been created to come up with a shared strategy for preserving legal information.[5] The LIPA approach is conceptually consistent with the strategy of creating a digital depository library system using cloud computing. Libraries commit to work together in the near term to retain a minimum set of physical copies of materials and over the longer term to develop standards for digital preservation. In a related effort, a nonprofit consortium called the Law Library Microform Consortium (LLMC) is pulling together a vast number of physical copies of our primary law to preserve them in multiple formats: in digital images, in high-quality microfilm, and in their hard-copy form in a salt mine beneath Kansas. Through efforts like LIPA and LLMC, legal information is being preserved systematically in ways that should allow reformatting over time, presuming that it is economically and technologically feasible.

Others are addressing the problem from the technological angle, such as those developing the promising "hub-and-spoke" model. This approach establishes a system that allows the data (the book itself) and the metadata (the catalog record, for instance) to be moved between a hub repository and others (the spokes) without losing any data in the transfer. The federally funded Hub and Spoke (HandS) project at the University of Illinois is an example of such a system. A hub-and-spoke architecture has, at its core, a common standardized information packaging system. This packaging system allows the hub and the spokes to communicate with one another intelligibly. Through this model, information can always be shared in a common format, via the hub. The spokes can, in turn, share information with one another via this system over time.

There are many variants on the hub-and-spokes theme. This heterogeneity is both a blessing and curse for the future of data preservation. Initiatives with names like Lots of Copies Keeps Stuff Safe (LOCKSS), the Northeast

. Document Conservation Center (NEDCC), the Unified Data Formats Registry (UDFR), PANDORA (Australia's web archive), Portico, and PRONOM (the National Archives of the UK's entrant into the field) all have promise. The technical details behind these systems matter less than the general point: we face not just a proliferation of data formats and the prospect of data rot over time, but also a proliferation of human efforts to solve the problem. It is great that so many people are working on this problem, but it is not so great that we are not yet working together in a way that will prove sustainable. These nonprofit preservation efforts need support from the commercial publishing sector to help drive a standard. The systems of information creation and preservation over time are too disconnected at the moment, which means there is too much risk of failure over the long run.

The technology problem of interop over time is a challenge to solve, but it may be human interoperability that needs to come to the rescue. The long-term issue is a problem of coordination of human efforts to ensure that reformatting happens regularly and consistently. It is essential that the system of digital preservation endures, even as the initial core group of collaborators retires or moves on to other endeavors in life. The interoperability issues at the technology and data layers have to be solved, but what is most important is that we build in interoperability at the human and institutional layers—that librarians, publishers, and the state work together on this common problem—for the long run.

I magine, for a moment, that we solve the interoperability issues, at each of the key layers of the interop model, for the preservation of books and journals and have a system that can work well for such formally published materials. But it would not solve all the problems of preserving knowledge because the transmission of human knowledge is becoming increasingly informal and increasingly digital. The authors and creators of knowledge are often amateurs. In a digital age, we call them bloggers, tweeters, remix video creators, podcasters. And they are e-mailers and texters and game producers and the gamers who play them. They (rather, we—it is all of us) create digital content outside the traditional confines of commercial and academic publishing. One might reasonably ask what we should do about all this digitally mediated human knowledge being created outside of peer-reviewed and commercially driven processes.

In one sense, the formal digital depository process for books and journals proposed in this chapter would also help preserve much of this informally produced knowledge. Some aspects of the system—such as the agreed-upon standards and the processes for updating formats over time—would ensure that there was a mechanism for updating files of many sorts without requiring that everyone have the exact hardware and software from the era when the knowledge was produced. This is one way in which the market might work: if someone cared enough about a work to reformat it, a way to pay to have it done would be assured.

But much would be left behind. Even the costly and ambitious proposal that we describe in this chapter would probably fail to preserve many aspects of works produced in a digital era. The writer and critic Nicholson Baker, for instance, has taken librarians to task for failing to consider the importance of preserving "dimensionality" in works so that future readers can experience them exactly as they were in their original forms of publication. We admit that Baker's version of complete preservation would be a far better one than the system we propose here, in many respects. But his approach would call for a level of commitment to preservation that is an extraordinarily far cry from where we are today.[6]

Our proposed system's imperfection—its lack of comprehensiveness—would in fact be a good thing, from at least one angle. Investing in the preservation of all the world's knowledge, regardless of its worth, would be a mistake. It comes as a surprise to many people that academic librarians often throw out books if they cannot give them away. Many librarians enjoy doing so. It is called "weeding," just like in a garden. (Librarians sometimes have to dump the books in recycling bins far from their patrons so as to avoid upsetting them. Faculty members at elite universities have been known to "dumpster dive" after cast-off books and to report, irate, to head librarians about the misdeeds of the collection development librarians, who were throwing out precious resources.)

Librarians are smarter than the dumpster-diving professors: they know that there is only so much information that we can afford, or even want, to keep around. Librarians know that we have to make choices about the quality and the value of information. It costs money, time, and attention to have information at hand. Collection development librarians make choices, every day, about what we ought to collect and what we ought to preserve. If we are honest with ourselves, not all of our bon mots are in fact so "bon." Much of the material we create belongs in the dustbins of history.

The job of preserving informal, born-digital materials requires librarians to have initiatives and skill sets that were not necessary in the past. Libraries have launched initiatives enabling professional librarians to choose which born-digital materials to preserve. A number of law libraries, for instance, have banded together to preserve certain web-based materials through an initiative called the Chesapeake Project. In this model, librarians determine which websites ought to be preserved as part of a shared collection. They use a series of common tools to preserve the content of the sites. They will still need to address the long-term interoperability issues. But initiatives such as the Chesapeake Project at least buy some time. And they have the added advantage of selectivity—though, in all likelihood, projects at this scale will prove to have been too selective. We will not have preserved enough of this informally produced knowledge to meet the needs of future scholars.

H istorians of the future ought to be jumping up and down right now, complaining about all the born-digital material not being preserved. Instead of diving in dumpsters and grumbling about throwing away redundant copies of old periodicals, researchers should be complaining that libraries and publishers and states have yet to solve the problem of born-digital archival materials in any way that can inspire confidence. For some, such as Nicholson Baker, even if we succeed in setting in motion the interoperable systems we propose, we will still not be doing nearly enough to preserve the world's knowledge.

The preservation of knowledge over long periods of time is too important for us not to address it with great care. We need to do so urgently.

There are some wonderful experiments under way, such as the work at the National Archives, the HandS project, and the Chesapeake Project. But we, as a society, need to rally collectively around this cause. We need to ensure that we are preserving more of what we ought to preserve and that we are able to forget what is sensibly (and forgivably) forgotten. The application of a sound theory of interop over time can help resolve the essential paradox of the digital era: our current tendency to preserve too much and too little human knowledge at once.

CHAPTER THIRTEEN

Architectures of the Future: Building a Better World

We, as societies, are rapidly building the information architectures of the future. As we do so, much will turn on how interconnected these systems are and how their interoperability is managed. For instance, interoperability is at the core of the fast-growing social web, of Facebook and Google and those whole swaths of life mediated by networked mobile devices. The principle of interoperability is proving essential as we move our health records into electronic formats in search of better care at lower costs. It makes possible the air traffic control system that keeps travelers far safer than they would be without it. Interoperability is central to the development of sustainable global marketplaces of ideas, goods, and services. Interoperability is a key to the long-term preservation of the world's knowledge and heritage.

Three new architectures on the horizon—cloud computing, the smart grid, and the Internet of Things—show us why it is vital and urgent that we get interop right at the levels of both theory and practice. These systems also serve as targets to aim for; each of these emerging systems is designed to be part of the solution of some of the most pressing problems humankind faces. For each, interop is a key design consideration. The degree to which they are interoperable will determine, in no small part, how effective the systems are. Without enough interop, these systems will not come into being. If they are made to be too highly interoperable, new problems could arise. Each of these three emergent architectures will enable us to create a better world if we strike the right balance between seamless interconnectedness on the one hand and extensive amounts of friction on the other.

Interoperability is an essential factor in the development of these three developing complex systems. The cloud provides an essential part of the computing infrastructure of the future; individual users, entrepreneurs, companies small and large, and governments around the world will rely on it. The smart grid has the potential to solve the energy crisis that we face from San Francisco to Beijing and in every city in between. The Internet of Things might well establish a highly interconnected, data-generating universe that opens up new spaces for creativity, innovation, and experimentation.

All of these architectures of the future demonstrate how important it is to get interoperability right at the theoretical level. They also show how hard it can be to make the right design choices when it comes to implementation at the four layers of the interoperability model. These future-oriented examples demonstrate the importance of breaking down a series of barriers, from the technical to the cultural, to the establishment of meaningful interconnectedness. Finally, these examples make plain the importance of collaboration, over extended periods of time, between and among private-sector and government actors for getting interop right.

C loud computing is the primary new infrastructure for computing and the Internet. In a cloud-computing environment, the basic functions of computing remain the same. We still use computers to share information with one another socially and professionally, to book travel, to purchase new books, and so forth. In most cases, we as consumers do not notice much of a difference when our experiences online are powered by cloud computing. But a new kind of magic is at work behind the scenes. The way in which these tasks are carried out is changing rapidly in a cloud-based environment. These technological differences may seem subtle, but they are in fact profound. They involve and in turn lead to much higher levels of interoperability in many respects.

Cloud computing is the delivery of computing as a service rather than as a product. The difference between the previous computing paradigm and cloud computing is the idea that shared resources (such as data) and software are provided to computers and other devices as a metered service over a network. Typically, that network is the Internet. In a world of cloud computing, one would never need to get a disk in the mail to put into a personal computer. In a cloud-based infrastructure, all the information, computer code, and processing power is located on the network. The personal computer—or smartphone, for that matter—functions mostly as a simple device to access what is happening online.[1]

Cloud computing is a familiar concept, at one level. Many people already rely upon Google's Gmail service for their e-mail, use LinkedIn and Facebook to manage their contacts, and share files with one another through Dropbox and YouSendIt. These services demonstrate the basic idea of cloud computing. Any user who has a web browser and a network connection can access information he or she has stored online and can process it using technology that is hosted in the cloud. Very little needs to be kept locally on a computer or a smartphone. The functions are similar, but the architecture that supports them is different.

Although these changes for ordinary computer users are important in their own right, cloud computing brings with it more substantial changes for businesses and governments and for the way they handle information. In the context of businesses and governments, a more formal definition of cloud computing helps illuminate the changes underway. According to the

US National Institute of Standards and Technology (NIST), cloud computing is a "model for enabling convenient, on demand network access to a shared pool of configurable computing systems (networks, servers, storage, and so forth) that can be rapidly provisioned and released with minimal management effort or service provider interaction."[2] This formal definition is useful because it focuses attention on the way professional technologists handle their systems. Just as consumers might choose to keep all their e-mail on Google's servers and rely upon Gmail as an e-mail program, IT managers can provide more extensive services to their users while avoiding costly, time-consuming processes associated with managing hardware locally and downloading software to every workstation.

Cloud computing does not entirely change what businesses and governments can do using computers, but it makes things much cheaper, faster, and more efficient. For example, the city of Miami relied upon cloud computing to improve its 311 nonemergency phone line, which citizens use to report issues such as potholes or missed garbage collections directly to the local authorities. Using a cloud computing infrastructure, Miami developed a web application that enables users to track service requests online and to see other requests that have been made in the area.[3] The difference that cloud computing has made in a case like Miami's is that the city did not have to develop all the functionality on its own, nor did it have to invest in expensive infrastructure to make this new service run. The city of Miami only had to build an application that established an interface to cloud-based services on behalf of its citizens.

Cloud computing is also making a big difference for small companies, which can now launch ambitious services that they never could before. A Canadian entrepreneur, for instance, had the idea of creating a new type of online bookstore, one that mimics the shelves of a brick-and-mortar store. In this new online store, book covers are organized in the traditional way, where one can get lost for hours browsing books as in a real bookstore. After a few months of work, the entrepreneur launched the new online store.[4] In the cases of this new online bookstore and Miami's pothole-reporting application, cloud computing's networked infrastructure made the deployment of ambitious projects possible on a time line, and at a cost, that was previously unthinkable. It was possible to create such systems before, but at costs that were prohibitive for many organizations.

As with most things in computing, matters get complicated very quickly. Cloud computing services come in many flavors, with acronyms such as SaaS (cloud software as a service), PaaS (cloud platform as a service), or IaaS (cloud infrastructure as a service). It is also possible to build both public and private clouds, as well as hybrids. The basic idea behind the fundamental shift in IT, however, is straightforward: in the same way that many of us are using web-based e-mail services such as Gmail or Yahoo! Mail, businesses can use cloud services to store and process their data and to develop and run applications. Businesses can also create entirely new services for their customers that connect to Gmail or Yahoo! Mail. For instance, software-as-a-service models allow for contacts in one program, such as Gmail, to be integrated with an online service like Salesforce, which salespeople use to track leads and monitor progress with customers.

The benefits of cloud computing are many—and they depend upon interoperability in order to be realized. For individuals, cloud-based services— ranging from Facebook (for social information) to Dropbox (for basic text files) and SoundCloud (for music)—are convenient, and they are often free to consumers. These services also provide a level of data security that would be hard, perhaps even cost-prohibitive, to achieve on a single personal laptop. For companies and governments, cloud computing is typically a matter of cost savings and time-to-market. Cloud computing offers the promise of low-cost, flexible access to scalable computing resources and enables institutions to outsource noncore activities to someone else's staff. The US federal government, for example, launched a landmark cloud computing initiative in 2010 that is expected to reduce its data center infrastructure expenditure by approximately 30 percent over the coming years.[5] If the systems involved do not have a high degree of interoperability at multiple levels, the government will not accomplish its cost-reduction goal.

Cloud computing is by its very nature all about interoperability. The entire point of cloud computing is that it allows for new levels of interconnectedness—at the network level, instead of at the level of personal computers. Systems that run in the cloud need to be able to interact seamlessly at the data and the technology layers. For instance, in order for a contacts system in Gmail to be able to connect to Salesforce's software in the cloud, these systems need to be made highly interoperable with one another, either directly or through an intermediary. In the previous computing environment, the systems involved were vastly less interoperable. The computers on which users stored their contacts would not be connected to the computers that stored the sales data, processing power, and software to run the programs that supported the sales operation. In the cloud-based environment, the level of interconnection across the board is substantially higher.

As a consequence of these heightened levels of interconnection, this new computing architecture also brings with it new challenges. The list of concerns is long. In a world of cloud computing, will we know in which country the data we are using are hosted and stored? Which government has jurisdiction over data in the cloud? Will our existing safeguards prove strong enough to protect confidential data in the cloud? Who owns data in the cloud? Who can access those data? What happens after data in the cloud are deleted? The list of issues related to data location, privacy, security, and ownership goes on and on. Accountability and liability are major questions: who will pay, or how will costs be apportioned among multiple players, if a highly interconnected system breaks down? The answers to these questions will be especially important when various companies work together in the cloud in the course of providing services to end users. Transparency is also a major consideration: how will computer users know who is handling their data, and where they are located in the world in geographic terms, at any given moment? In this new, complex environment, policy makers worry about tracking information, not to mention liability, among various different clouds, which may be held in a mix of hands and across multiple jurisdictions.

To delve into the specifics of how to create and manage interoperability in the cloud, let's start with cloud-based social networking sites like Facebook, Orkut, LinkedIn, Ning, or Xing (a social network of businesspeople in Europe, with over 10 million members). These services are familiar ways that consumers use cloud computing to access personal and social information. Technical interoperability in this type of communication service typically has two dimensions: vertical interoperability (interoperability within a single platform) and horizontal interoperability (interoperability across different platforms). Despite being hosted in the cloud, many social networking sites are constructed to be deliberately noninteroperable along the horizontal dimension. In other words, they are designed not to work together with competing social networking sites.

There is very little horizontal interoperability across social network sites, which are holding important personal information in a series of private clouds. There is no simple way to move your account data—a feature called "data portability"—from one social networking site to another. However much you might want to, you simply cannot move your Facebook friends to Twitter or from Twitter to LinkedIn, and so forth. You would have to download each individual photo, link, and note and then upload them all over again. The logic behind these interoperability barriers goes back to the phenomenon of network effects. The greatest asset of companies such as Facebook and Twitter is the breadth of their user base and the quality of the "social graph" that this broad user base comprises. The more people use a particular service, the more likely others are to join. From a business perspective, it makes a lot of sense for these companies to lock in users by making it attractive to join but difficult to leave and start a new profile with a competitor.

The kind of interoperability we see in the cloud-based environments of Facebook, Orkut, Twitter, and LinkedIn is vertical, not horizontal. These services have built in a great deal of interoperability within the services they control but fairly little across competing services. For instance, they excel at implementing new ways for users to take content from other sources and rebroadcast it through their own services. Facebook recognizes when users include a website or video in status updates and pulls up thumb-nails or even onsite portals so viewers can watch the video without leaving the page they are on. Twitter recognizes links and encourages shortening the links to fit its 140-character limit. Similarly, these companies want to maximize their user base and convenience by making themselves broadly accessible, and so they develop apps that allow users to access their services on mobile phones, some even through SMS platforms. This interconnection is made much simpler and more extensive by the fact that these services reside entirely in the cloud.

Cloud-based services also support innovative ways to improve health care through better access to health information. Interoperability made possible by the cloud is at the heart of these improvements. For instance, Microsoft's HealthVault enables patients and their doctors to do a better job managing diseases like diabetes and heart disease. Patients at the Cleveland Clinic (a nonprofit medical center that performs care, research, and education functions) use at-home monitoring devices to measure data on glucose levels or heart rates; the data are uploaded to HealthVault and are then immediately available to the patients' health care providers, allowing them to follow the progress of their patients remotely and to draw on more comprehensive and accurate data when meeting with the patient at the clinic.[6] In a precloud environment, this exchange of information would be much slower and more expensive. The vertical interoperability within HealthVault and related services allows people in separate locations to work together on a common task, such caring for a patient's health. Neither the patient nor the doctor needs expensive computing equipment or software on their local machines; they just need a way to access the Internet. These cloud-based technologies have been well received by patients and hold much promise for improving medical treatment.[7]

These examples of health care and social networking show the effects of moving computing power and information to the cloud. In both cases, the cloud-based approach leads to a higher degree of interconnection among people and data. These examples illustrate how high levels of vertical interoperability can greatly benefit end users by linking together different devices, databases, and ultimately people. These examples also illustrate that interoperability in the cloud cannot be taken for granted. Interop in the cloud needs to be created by design and managed with care.

Cloud computing is just emerging as an important computing paradigm. The emerging challenges associated with cloud computing are close variants of the challenges we have explored throughout this book, yet they are heightened by the extensive degree of interoperability. For instance, cloud-computing service providers sometimes have strong business incentives to limit horizontal interoperability across platforms or services, much as Apple initially limited horizontal interoperability with its music services. Complex technical issues also limit certain kinds of interop. Proprietary data formats used by different types of cloud service providers may limit the extent to which data can be rendered useful from one part of the cloud to another. The lack of open and standardized infrastructure formats for data puts limits on data portability even in cases where services might choose to work together. Noninteroperable contracts (known as service level agreements) among cloud providers can also work against portability by setting different parameters for how an end user may interact with a cloud service and how that user's data ownership rights are governed.

Many of the interoperability challenges listed here—especially the technical and contractual issues—are best resolved by industry players. Current industry-led cloud initiatives, such as those hosted by the World Economic Forum and the Aspen Institute, are designed to resolve these exact problems by bringing together the big players in the cloud-service industry. This collaborative approach may well lead to new interoperability standards and common protocols that will help take advantage of the best aspects of cloud computing while mitigating the problems associated with it.

As in other interop settings, there is also an important role for governments to play. States can do more than merely stand by to intervene if market forces fail to resolve some of the important interop issues. Governments can serve as conveners and facilitators of standards-setting initiatives. NIST's important work—on definitional challenges, research, development of use cases, and reference architecture development—nicely illustrates how this government role can be played to benefit various stakeholders, including consumers. The US federal cloud-computing strategy, an important statement of the power of government procurement (which allots $20 billion in federal IT spending to the cloud), will shape the cloud-computing landscape. And finally, there is one challenge that only governments can resolve. An additional layer of particularly persistent interoperability problems in the cloud—where data cross borders—stems from divergent national laws and regulations. Governments in the digital age must grapple with the essential task of ensuring that these institutions work together better than they do today.

T he electrical grid is remarkably similar to cloud computing in terms of how it works. Both are, by design, highly interoperable complex systems that enable ordinary people and businesses alike to draw upon common resources to carry out everyday tasks. On the electrical grid, consumers demand and receive power without needing to understand the devices or infrastructure that deliver what they need. Interoperability enables the electrical grid to function seamlessly for vast numbers of consumers, who do not have to call the power company every time they want more electricity to flow to their home or business. Interop is also at the core of the next generation of electrical grids, which are expected to be much "smarter" than the current systems.

The electrical grid is among the most significant engineering achievements of the twentieth century. This complex network—power plants, transmission lines, and other components—that enables the generation of electricity and the transmission, distribution, and control of electrical power is itself a marvel of interoperability. The next phase of development of the grid is probably the construction of the "smart grid," which would be a vastly more efficient way of allocating energy and a boon to the environment. The development of this new layer to the grid involves creating new ways to share information between and among parties about the flow of power.

The development of the smart grid is one of the most important interoperability problems on the horizon. In addition to linking virtually all homes and buildings into a power network, the smart grid would establish high levels of interconnectedness in the data about energy demand and consumption. These new forms of interoperability bring with them the promise that the smart grid will operate more efficiently than the existing power grid, but they also bring new problems, such as concerns about privacy and security.

Large parts of today's power grid are based on design principles and implementation choices that grew out of the first electrical networks and the technology available in 1900. Since then, the world has changed dramatically. The population in need of electricity, on a global basis, has grown exponentially. New sources of demand—especially due to the increasingly widespread use of computing and other electronic devices, such as televisions and radios—have put additional pressure on the grid. As grids have become increasingly large, interconnected, and international, their vulnerability has increased as well. The massive blackouts affecting millions of people in the United States over the past decade are important reminders of the enormous pressure that modern life puts on our largely outdated electrical infrastructure, even in the most highly developed nations. Many additional challenges, including security threats, lurk just underneath the surface.

Power companies and their regulators have launched large-scale initiatives to modernize today's power grids and to make them more reliable, efficient, and safe. Government regulators, as well as those who provide power around the world, are thinking hard about the capabilities that a modern grid must have. Among the top requirements is the ability of the network to "heal itself." The idea is that the grid can be configured so that it can deal automatically with problems such as power outages or service disruptions. Information from grid usage can be used to nudge consumers toward better behavior, so as to save energy and to allow the grid to operate more efficiently overall.

The smart grid is in its infancy as a technology. It is not widely deployed in a way that is obvious to consumers, but a great deal of planning and development is underway. The smart grid promises to deliver electricity from suppliers to consumers, but it will also offer built-in digital technologies to facilitate communication between them. Today's digital communication network makes it possible for sensing, measurement, and control devices to be made to interoperate. These devices can collect and pass information about the condition of the grid among themselves, allowing the grid to respond dynamically to events that occur anywhere in the power generation, distribution, and demand chain. The smart grid can adjust the power flow in response to changes in the environment—for instance, by throttling down what each home or office can demand in very hot weather to avoid brownouts. The smart grid can also alter demand in positive ways. Consumers can make better choices about what they really need at peak times. The smart grid allows for dynamic pricing during peak periods of usage, making consumers smarter about the real costs of energy consumption. The system can also act dynamically to prevent systemic failure by temporarily shutting down a distribution line at crucial moments.

New, interoperable appliances can help consumers act in ways that protect the environment through energy conservation. The deployment and integration of smart consumer appliances and devices such as smart meters and smart thermostats, along with automated control of equipment, will help empower consumers to respond to changes in the grid and adjust their behavior accordingly. Before the smart grid came along, this option was only available to very large energy consumers—such as providers of cloud-computing services who need extraordinary amounts of power to run their massive systems. Under the new paradigm, users can allow the smart grid to turn off certain appliances, such as dishwashers or washing machines, during peak times to reduce demand and cut costs. The smart grid also enables decentralized sources of power—for instance, energy generated via solar panels on a house—to be fed back into the system. This feature becomes particularly useful for emerging building prototypes for structures that generate more energy than they use.[8]

The smart grid is not a substitute for the traditional power grid. It is an overlay, built on top of the ordinary electrical grid, made up of highly complex communication equipment and sophisticated metering system. Think of the smart grid as an "energy Internet" that integrates a number of different technologies and functions into one network of networks. Each of the technologies that is part of the smart grid (ranging from smart sensors to improved grid-level storage technologies) has positive effects on its own.

But when these components can all work together in a coordinated way, using all layers of interoperability, they create significant efficiency gains and will become a pillar of any future solution to the energy and climate crises we face. In some sense, interoperability on the smart grid *is* the smart grid.

The interop problems associated with the emerging smart grid are many and complex, and for the most part they are far from resolved. Thus far, only pieces of what will eventually be the smart grid exist. Many of the smart appliances that will support the grid are still just on engineers' drawing boards. Those that are in production are not in widespread use. There are still major capital costs to be borne by energy companies to support these appliances and the related information networks. In addition, all the necessary technologies and the ways in which they are supposed to interact across the different layers of the smart grid are today insufficiently standardized.

The first problem associated with interoperability and the smart grid is the basic fact that the smart grid does not yet exist. It is a fruitful example of an interop problem because it is on the drawing board, calls for high degrees of interconnectedness, but has not yet been built out in full. The levels of interoperability still need to be set, and many different people would like to have a say about them—certainly, the companies building the smart grid and their regulators have a role, but so too do consumers and those in civil society who look out for privacy concerns and issues related to public security. The whole set of problems associated with high levels of interoperability will arise. How can we ensure that data about our private activities in the home are not shared with the wrong people? How can we keep the smart grid from being hacked by terrorists? How can we ensure that technology that works in 2015 will still work in 2025, or is at least flexible enough to adapt, avoiding the problem of lock-in? How can companies and governments build smart grids that allow for—better yet, encourage—diversity and innovation?

The enormous scale of the smart grid poses special challenges for interoperability. The number of players involved is vast: customers (which means virtually everyone in a given society who is on the grid); utilities;

equipment designers and manufacturers; local, state, and national governments; and environmental groups have a stake in the outcome of the standardization process. Add in the international dimension, and the challenges multiply. In addition, a version of network effects works against interoperability. Efficient peak pricing is such an example, where the benefits only emerge when a certain number of people buy into the system and purchase smart measurement devices that report back accurate data. The delayed return on investment at the outset makes people reluctant to invest in this type of technology, which slows down the adoption of interoperable smart grid technology.

Once established, there will be drawbacks to interoperability on the smart grid. One major risk of broad interoperability based on standards-setting is technological lock-in. Standards-setters need to ensure that the rules established at the beginning of the smart grid's existence are stable enough to support development of the smart grid but malleable enough to support innovation over time. Security is a second potential drawback. A highly interoperable smart grid is more complex than the less-networked grid of the past, offering more points of vulnerability. Standards need to be set that will address the cybersecurity risks presented by full-scale implementation of the smart grid. And privacy concerns associated with the smart grid are paramount. In the extreme case, any switch of the light from on to off and vice versa would be tracked, reported, and analyzed. These data, if not properly safeguarded, could be misused by marketing firms, burglars, stalkers, and others who would do harm to users of the smart grid.

Whether in the United States, Europe, or Asia, governments are the major drivers of interoperability on the smart grid. They acknowledge the need and desire to create the smart grid. Everyone knows that we cannot enjoy the benefits promised by the smart grid without getting multiple aspects of interoperability right. But optimal levels of interoperability, across so many dimensions and involving so many actors, are not easy to achieve. These government actors also see the range of possible drawbacks associated with increasing the degree of interconnection on the grid.

The development of the smart grid in the United States is guided by a public-private effort with real promise. NIST, as in the case of cloud computing, is helping guide industrial development of this next-generation system. NIST is leading an industry-wide, collaborative, and, so far, constructive standards-setting process for the smart grid. A federal law—the Energy Independence and Security Act of 2007—tasked NIST with developing a framework for interoperability on the smart grid, and Congress has funded NIST to carry out this important work. In cooperation with the US Department of Energy, NIST has identified the key issues associated with interoperability on the smart grid. The primary goal is to establish the right standards to facilitate optimal levels of interoperability on the emerging smart grid. Smaller initiatives target key issues, such as cybersecurity and privacy, that must be addressed before the smart grid connects all our homes and businesses in new ways. The group has identified over a dozen standards as priorities, including standards for smart meter upgrades, common specifications for price and product definition, energy use information, and precision time synchronization.[9] The group has come to early agreement on the way information will be used to communicate between the utilities and the consumer and the way information is to be organized.[10] This collaborative, design-oriented model holds a great deal of promise for getting interoperability right as the smart grid comes online at scale in the United States.

This development of the smart grid illustrates why interop by design is so crucial. The drawbacks of smart grid interoperability have to be taken seriously and considered carefully at the outset of the process. These potential or actual costs have to be weighed against the benefits of what interoperability enables. In the case of the smart grid, interoperability is a constitutive principle; it is essential for the system to work. The smart grid is a pure interoperability case: the enormous benefits that the smart grid offers are the benefits of interoperability itself. Higher levels of interoperability can improve the reliability and efficiency of the electrical grid, reduce the price of electricity, create a platform on which new products and services can be developed, and promote environmental quality and renewable energies. On balance, the question is not whether or not we should have smart-grid interoperability but, rather, how we can work together to overcome the remaining barriers and deal proactively and responsibly with the potential drawbacks.

A third architecture of the future that depends heavily on interoperability is the Internet of Things. The IoT, as it is called, is an emerging information network that, counterintuitively, has to do with physical objects. The basic idea behind the IoT is that virtually every physical item—a razor blade, a bottle of water, a radiator, a chair, a car—can be turned into a type of tiny computer (a "smart object") and be connected to the Internet. Of the three emerging examples that we describe in this chapter, the IoT is the most speculative and the least certain to develop in a predictable fashion. The future of the IoT depends to a large extent on the question of whether we will be able to overcome the various interoperability barriers at many layers. The IoT is also a development that is more controversial than either cloud computing or the smart grid: it is not clear to many people whether in fact it should be built at all.

The benefits of an IoT could be substantial and widespread, but they are speculative enough to be hard to see. The world is awash in data. A networked universe of things, each connected to the Internet, would enable us to collect, aggregate, analyze, and use these data in unprecedented ways. An IoT could help improve our lives as patients; could help companies, markets, and governments work more efficiently; and, most important, could let us begin to address some of the most pressing societal challenges we face, including the efficient allocation of natural resources such as energy and water.

Interoperability is the DNA of the IoT. Interoperability issues are written all over it, ranging from the purely technical, as in the case of RFID standards, to the institutional, such as adequate legal safeguards for privacy. The approaches used to overcome these interoperability problems has to take into account drawbacks such as information overload, threats to privacy, and security concerns.

The IoT is made up of a universe of smart objects. The idea of smart things is not new, but it has only recently become possible to produce extremely small and low-cost networked computers that can be merged with physical things. Although the grand vision of the IoT, in which billions of things are connected with each other, is still a dream (or, for many people, a nightmare), the first instances of the IoT are beginning to come into view.

The tensions inherent in the IoT as a concept—its promise as well as the fears to which it gives rise—are revealed through an examination of a series of experiments. One such experiment is an unusual building in California. A team of visionary researchers at the Mobile and Environmental Media Lab at the University of Southern California asked, If a building could talk, what would it say? They wondered how a building might "feel" about the comings and goings of people, whether it could be affected by their moods and desires, and what kind of relationship it could have with its occupants if it could communicate with them.[11] The lab team framed the questions as an experimental design project, created a vision of a building, and identified a set of innovative technologies that could give answers to these questions. This is the birth of what is known as the Million Story Building project.

Using an actual building at the School of Cinematic Arts as a test environment, the lab team designed a series of location-specific interactions in the built environment and created an interface to the building by using mobile phones, sensor networks, and software applications. Through these technologies, the students, faculty, and staff of the school can, in effect, *interact* with the building on a daily basis and, in some sense, develop a relationship with it. The idea is to learn about the people in the building and what they are doing there. The building creates user profiles by aggregating data about its inhabitants, learning names, locations, and activities; in turn, the building can offer back to its visitors tailored information according to their perceived interests. Using movie clips, photos of different areas of the building, and other digital materials, the building introduces itself to its visitors via digital technologies—think of interactive applications on a smartphone that guide a user through the smart building. The effect is that visitors encounter a gamelike environment as they use the building. People are asked to complete more difficult tasks within the building over time, much as gamers perform harder and harder quests in video games. As inhabitants interact with the building and provide information

about themselves and what they are doing, the building records these activities as a digital archive—a living history of the new building.[12]

The concerns associated with the Million Story Building project are as
easy to see as its attractions. Smart buildings, enabled by the IoT architecture, can help visitors find their way and move around efficiently, can perhaps even have personalized cups of coffee waiting for them at a kiosk near
their workstations or classrooms. Perhaps smart buildings will interact with
increasingly precise fitness applications, such as Fitbit (think of a networked pedometer for the social web), to help visitors stay fit as they go
about their daily routines. The benefits for social and architectural historians are also plain: wouldn't we love to know how people in antiquity made
their way through their built environments?

The same technologies that make these benefits possible—primarily,
localized sensor-based networks—also give rise to Panopticonesque fears.
We have already lost much of our privacy by recording our social lives online; the IoT might well lead to a similar deterioration of individual privacy
in physical space as well.

Jails are experimenting with early implementations of the IoT.[13] A county
jail in the United States has started using extensive RFID technology to
track detainees and guards, enabling jail officials to understand better the
interactions among them. On visiting a cell, a guard first scans a tag on the
doorframe, which records his presence at the cell. The guard then scans an
RFID wristband on the prisoner, which records the prisoner's identity
along with the date and time. Additional information, including the reason
for the visit, is recorded as well. The benefits associated with this experiment might extend beyond the obvious security implications. If theory
holds, the effect of recording these real-space interactions may be to encourage more appropriate interactions on the part of both the guard and
the detainee. But the well-known problems raised with Jeremy Bentham's
Panopticon could be raised about this IoT-powered prison experiment, too.

Schools, too, may soon turn into highly networked computing systems.
A research project is offering an East Coast school funding to experiment
with RFID technology.[14] Early ideas for this project include simple mea-

sures, such as tracking school property—laptops and books in the library, for instance. More extreme versions include the tracking of students, for example, by giving them RFID-equipped backpacks. This measure could enable parents, school officials, or police to locate students quickly in case of an emergency. Some parents like the idea of their child's backpacks being trackable, much as they like the GPS functionality in the smartphones they give their kids at ever-younger ages. But these experiments, too, trigger serious privacy concerns and well-placed worries that children are losing the ability to grow and thrive while not being tracked at every moment.

The benefits of highly networked tracking equipment are easier to see in the case of health care than in the case of schools. Hospitals, for instance, are increasingly turning to IoT-powered technology to improve patient monitoring, automated medication, and the preservation of patient data. In hospitals in some developing countries, RFID tags for infants are used to prevent baby thefts. These examples also show the connection between two major interop stories: the IoT and the electronic health records example we explored in depth in Chapter 11.

IoT applications can serve trivial purposes as well as profound. The IoT has recently made its way into the coffee shop, for instance. The SMUG is a smart mug, outfitted with an RFID chip that enables personalized ordering and fast purchasing. SMUGs allow customers to communicate their coffee preferences and payment information to their favorite coffee shops—a step further than the Starbucks smartphone payment initiative that we mentioned in Chapter 3.

If you have made it this far into the book, you can no doubt see the extent to which the development of an Internet of Things depends on interoperability. And, in turn, as the IoT grows, the degree of interconnectedness and interoperability will increase. Unlike the Internet, however, the IoT brings the effects of high degrees of interoperability out of the digital realm and into the physical.[15]

It is not obvious how to build the IoT at scale, even if the fears associated with it were set aside. The physical world has constraints not present in the digital. In the context of the IoT, we once again have to contend seriously with the geographic distances among trillions of potentially relevant physical objects. The characteristics of materials once again become relevant. And the topography of the surrounding environment again serves as a major constraint. This enormous diversity is a real challenge for establishing the IoT with high levels of interoperability, which is invariably context-specific.

These physical constraints are essential to the puzzle because the IoT requires object-to-object communication. Physical objects will need to transmit information, and many of them need to be able to "listen" and understand transmissions. In the IoT context, technical interoperability means that a signal can get from physical object A to physical object B. Semantic interoperability means that A (alarm clock) and B (coffee machine) can understand each other, that these objects "speak the same language." Except for the underlying wireless network systems that carry these signals, no single global standard has emerged that regulates interoperability—either technical or semantic—comprehensively.

Interop is hard to accomplish for the IoT at the technology and data layers, but it is even harder at the human and institutional layers of our model. Imagine the problems that occur if two hospitals, within a reasonable distance of each other, run different versions of the IoT, from different vendors, to track aspects of their patients' care. Even if each internal hospital system works seamlessly from a technical perspective, an interop gap will persist between the two organizations that will prevent them from working together in the most effective ways to help patients who move from one to the other. Even if the technological systems, as well as the organizational structures and processes, could be patched together in some ways, different cultural norms, legal requirements, and other higher-level interoperability barriers may stand between the two hospitals, their employees, and their patients.

These many challenges can be overcome, especially where a profit motive helps provide a driving force. The IoT is coming into being most quickly in the context of large-scale businesses that stand to benefit from tracking physical objects through networked technologies. The IoT is slowly but steadily coming into being in some industries, such as retail and consumer goods. Here, certain emerging de facto standards—for instance, the EPCglobal standards, which are helping standardize item-level tagging—have been established for key components of the IoT system, especially with respect to RFID technology. The use of these standards will probably expand to related industries, including the textile and pharmaceutical industries.

The development of the IoT links back to the discussion of systemic efficiencies in Chapter 7. There, we discussed how bar codes have been used to improve inventory management, logistics, and accounting processes. Large retail companies, such as Walmart, have pushed their top suppliers to use RFID tags on all cases and pallets of consumer goods shipped to its distribution centers and stores. RFID technology has increased efficiencies throughout its supply chain. For instance, Walmart has been able to restock RFID-tagged items three times as fast as nontagged items. Walmart's use of RFIDs is an early example of the IoT serving a constructive, demonstrative purpose. Similar uses of basic IoT technologies across organizations and units have been reported in other industries, including the automotive industry, where RFID technology speeds up vehicle pickup and improves customer service.

And yet despite these early successes, the vision of the ubiquitous Internet of Things—a system that connects large parts of the physical world with the digital—remains primarily on drawing boards and in computer science laboratories. Some people may prefer it to remain there, merely a concept for experimentation in universities and corporate R & D facilities. The success or failure of the IoT in the long run, as well as its desirability, will depend largely on how interop is established and maintained.

I nteroperability plays a crucial role as an enabler of these three emerging architectures of the future—cloud computing, the smart grid, and the Internet of Things. Interoperability in the cloud is essential from the user's perspective. The degree and the nature of interop are among the key factors determining whether cloud-based technologies will be adopted or distrusted in the long run. In the case of the smart grid, interoperability is absolutely essential; in some sense, interoperability is the DNA of the new grid. And the vision and practice of the Internet of Things cannot even be contemplated without positing high degrees of interconnectedness among things and between physical and digital space.

Interoperability issues arise at all four levels of our layer model in these three emergent examples. The interoperability challenges at the technical level are significant across all three architectures; they range from issues of data formats to intricate aspects of semantic interoperability. But technical interoperability is not the only serious challenge. In all three cases, the way people and institutions develop and use the interoperability in these systems is just as important as how the data are designed to flow within and across them. Each of these cases also poses legal and policy issues, especially related to privacy and security, as more and more data flow across the boundaries of states in ways that are hard to track and manage.

Although each of these three emerging architectures raises problems, as a matter of substance, the potential benefits of increased interoperability in each case should ultimately outweigh the drawbacks. That is certainly true in the case of cloud computing and the smart grid; it is less obvious in the case of the IoT, which is both harder to envision and more controversial on its face. In the near future, societies will need to focus on how to manage the costs and benefits of the interop that is inevitably part of each of these three systems.

No matter what complex system we decide to build next to make this world a better place, it will require a shared commitment to increasing the interoperability of our systems, our institutions, and ourselves in productive ways. Whether at the technology, data, human, or institutional level, the optimal degree of interop will emerge in these three cases, and in others we can hardly imagine, only as a result of a massive collaborative effort. A broad group of stakeholders, from both the private and public sectors, will need to work together strategically, in good faith, over a long period of time to get interop right. As a matter of process, the principal drawback of this type of collaboration is that these processes take more time than ad hoc, private-sector innovation ordinarily does on its own. And these processes are hard to pull off: they require deep trust, ongoing commitment to active engagement and openness, and a willingness by participants to set aside short-term gains in favor of shared long-term systemic improvements. The benefits of such large-scale collaboration, though, far outweigh these costs: stability as new systems come on line; efficiency and other immediate benefits for consumers and businesses alike, with well-managed downside risks; and sustained innovation in new emerging systems.

The Payoff of Interop as Theory

How are we to manage the unprecedented degree of interconnectivity that has been created between and among people and systems in the digital age? This is one of the most significant questions of our age. Much depends on our ability to maximize the benefits of this unparalleled and growing level of connection and information flow while minimizing its potential risks. We need to get interop right as a matter of public policy, as we address big issues like sustainability and climate change. Interop is also important in the private sector as a matter of strategy, in terms of helping businesses thrive and innovate. The theory developed in this book is designed to help consumers, business leaders, policy makers, and the public at large to make more informed—and ultimately, better—decisions about the ideal level of interconnectivity among complex systems and their components, about what we want to get out of interoperability, and about the breakwaters that should be put in place to make sure it stays at the optimal level.

The theory of interoperability outlined here can be used in four ways: first, as a framing device and an organizing principle—in essence, as high-level theory; second, as a description, to guide us in our understanding of certain phenomena, mostly to do with information and technology, in the age in which we live; third, as an effort to predict what the future holds and what debates will surround the subject of interoperability in years to come; and finally, as a normative device, one that should drive and inform the kinds of decisions policy makers ought to make in order to lead to the kind of good societies in which we all wish to live.

INTEROP AS HIGH-LEVEL THEORY

The theory of interop that we develop and test throughout this book draws together a series of seemingly unrelated events, innovations, and themes in such a way as to establish unexpected and revealing patterns. What, for instance, do the global economic crisis that started in 2008, health care reform, global climate change, and the emergence of the social web and cloud computing have in common? All have interoperability at or near their core, what makes them possible and what can make them dangerous. The study of interop helps us see the promise and the perils of highly interconnected systems in our increasingly globalized economy through the similarities and differences among these widely ranging examples.

As a theoretical framework, the study of interop sheds light on what tends to go right and what can go wrong with complex systems that rely upon a constant exchange of information, most commonly mediated by digital and networked technologies. Although some of the interop stories included in this book—such as the evolution of emergency systems, shipping containers, and bar codes—predate today's digital era, they have important relevance for interop in the current age. The implications of this theory of interop are highly relevant for the next generation of complex systems. After all, it was not possible for information to flow as quickly or as consistently across organizational and national boundaries even a few decades ago. Nor have people and materials been nearly as mobile and interconnected as they are today.

One of the key insights offered by interop theory is the degree to which the proper functioning of systems that seem to be predominantly technical in nature—such as air traffic control systems, cloud computing, or the smart grid—depends on how well human beings and institutions can work together. Over the past decade, much thought and money have been spent making information and communication technologies more robust and improving the systems that rely on them. It is crucial that we advance our technological know-how and practices to ensure that our data are safe and our privacy protected. But the theory of interop also highlights that we have to think equally hard about the appropriate design of the fragile interfaces where technology, data, human, and institutional layers intersect if we want to harness the benefits of the unprecedented interconnectivity in the future. Examples such as emergency communications and health care information teach important lessons about what has worked and what has not.

INTEROP AS DESCRIPTION

Interoperability research does not only lead to an abstract theory; it also helps at a precise, descriptive level. The careful study of interoperability helps explain specific phenomena in a complex world. An understanding of how interoperability functions in the context of case studies reveals much about what makes complex systems work well and what leads to their failure. Our methodology has been to explore case studies where we imagine interoperability might be part of the magic behind a system's functioning, for good or for ill. These case studies have taken us from the worlds of information technology, commerce, and trade to health care, emergency response, and the related fields of energy and environment. These case studies are posted freely on the web, at http://cyber.law.harvard.edu/interoperability, for anyone to read. These are the raw data and collected stories that we have worked from in the pages of this book; we have woven these narratives into the frame of our argument. They also stand alone as rich descriptions of how complex systems function and of where they can break down.

These case studies describe connections that are hard to see on the surface but that are essential to the functioning of our complex world. A look beyond the surface of everyday phenomena—such as digital music, bar codes on products, instant messaging, and shipment containers (the boxes in which goods tend to flow around the world, on large ships and on trains)—encounters the hidden links and information channels among systems, components, and applications. It also discovers how much their capacity to work together depends on a complex set of choices, made over a long period of time, by a large number of players. These players have typically included technologists, consumers, companies, legislators, courts, and others. To make things more complicated, many of these decisions have been made in an ad hoc, decentralized fashion—certainly without any grand interop plan to guide the way. Given this decision-making process, it is surprising how well many of today's systems work together and how interoperable our world has become. At the same time, many of the case studies also illustrate how hard it is to undo bad decisions of the past. The legacy problem and the problem of path dependency (which we observed especially in the library and e-health contexts) are reminders of how important it is to think about interoperability in a proactive, strategic fashion.

Interop helps us understand issues related to globalization and how our cultures differ from one another. A global perspective, as we look forward, can help expose culturally specific approaches to interoperability. China, for instance, with its enormous market size, has a particular set of strategic interests with regard to interop. Chinese government and private companies are developing independent standards for certain information and communication technologies outside the realm of the international standardization organizations described in this book. Chinese officials have seen the development of their own standards as a matter of potential competitive advantage, both in security and in the marketplace. Officials in the United States are beginning to see standardization and interoperability issues in a similar light.

Such diverging regional interop approaches are also visible in a comparison of everyday experiences. Consider, for instance, the dissimilar ways in which we in Western countries and our friends in Asia deal with different electrical plugs. In China, the solution to this annoying interop problem is not an adapter but, rather, a pragmatic, multiplug design of the power outlet itself, built into the wall. Or take the example of a contactless, interoperable smart card, called Suica (the Super Urban Intelligent Card), that is used to pay the fare on trains in Japan. This nicely designed card works outside of trains, too; the Suica is increasingly accepted as a form of e-money for purchases in stores, at kiosks, and in taxis. Meanwhile, in the United States, we carry around wallets stuffed with different credit cards, swipe cards to allow us various forms of access, and separate customer loyalty cards from our drug store, our grocery store, and the place where we get our coffee in the morning.

These examples from Asia suggest another lesson from our research: interoperability, in virtually every context we have studied, is in constant flux and is occurring at differing rates around the globe. Rapid technological progress combined with highly dynamic market forces will continue to create new interoperability challenges and at the same time change the character of old problems. But the problem side of the equation is not the only thing in flux. The ways in which we address interoperability challenges may change over time as well, because we will learn from our own successes or failures and will also be inspired by different approaches from other parts of the world.

This theory and these case studies may be most immediately relevant to those who work in the industries and areas examined in the specific cases, such as computing and the web, libraries, and health care information systems. The implications are easiest to see in the context of information technology companies. The importance of an interoperability strategy is obvious to those who work at Apple, IBM, Microsoft, or Oracle, in the high-tech world. Increasingly, the next generation of big information technology companies are betting even more on strategies of interoperability: Facebook, Google, and Twitter are all building enormous businesses by developing, and sharing wide access to, highly interoperable platforms. The same is true of companies all around the world, many in Europe and others in the fast-growing markets of East Asia.

But these issues are highly relevant to policy makers and consumers, too. The job of setting policy in the digital era increasingly calls for a deep understanding of interoperability and how it affects a broad range of legal and policy outcomes. It is an issue of competitiveness and of national security. And for consumers, the level of interop that people demand has a powerful effect on the decisions companies make as they design their products and services. Higher levels of interoperability can be great for consumers in terms of convenience, but it can also pose risks for security and privacy, as we have seen in the cases of Google's Buzz and Facebook's Beacon products.

INTEROP AS PREDICTION

Interoperability theory helps company executives and government policy makers by enabling them to make better predictions. The study of interop helps decision makers look ahead as they try to anticipate the results of their actions today. For instance, a large technology company may want to know whether it makes more sense to allow free access to and connection with core systems, opening them up to other developers (as Twitter and Facebook have done on the social web), or whether traditional strategies of exclusion are a better way to go. In the online world at least, the increasingly common answer seems to be that high levels of interoperability lead to better results for individual companies, for the industry at large, and for consumers.

But a well-designed interop strategy, as we have seen time and again, must also get the degree of interoperability right. It is essential to realize that high levels of interoperability can lead to further problems, often related to security and privacy, homogeneity, and lock-in. It is important to craft interop strategies that take advantage of what we know to be the major advantages of highly interconnected systems while working hard to design systems that mitigate its several potential downsides. Interop theory can help guide this design process.

Smart interop strategies adopted by tech companies, as well as sound interop choices made by users and regulators, will help harness the benefits of digital interconnectivity while avoiding its risks. But the most challenging interop problems often stem from the sheer complexity of the systems we want to make work together. For instance, it is very hard to envision what a successful interoperability strategy for the next generation of air traffic control systems will or should look like, because there are so many stakeholders around the world and so many different technologies involved. The same is true of international financial markets, where it is very hard to model the effects of the most highly interconnected systems and the most complex financial instruments. Viewed from this angle, our studies highlight the urgency and importance of sound interop strategies design to handle complexity at a global scale. Our theory demonstrates how users, companies, and governments should expect to come up against limits of how effectively we can predict outcomes in the most highly interoperable, complex environments—a major trade-off that we must realize we are making as we continue the process of deep interconnection.

INTEROP AS A NORMATIVE MATTER

Finally, the close study of interop helps determine what we, as societies, *ought* to do in certain circumstances. The study of interop can inform decision making about what the most promising approach might be to any given new interop problem. Interop theory helps us consider how we might solve the problems that we expect to face in the near future. The health care debate and the need to preserve human knowledge in a digital era, for instance, are two pressing issues that will require governments, companies, and consumers to have a firm understanding of interop issues. The emerging architectures of cloud computing, the smart grid, and the Internet of Things also present intricate interop problems that we, as societies, will need to address.

At a granular level, this emerging theory of interoperability provides a framework for sound interop policy making and puts forth a process-oriented model for policy makers who are seeking to address interoperability problems that have arisen or are likely to arise. Most of the cases we have examined here are not straightforward instances of clear lawmaking; they tend to involve cultural and societal factors that shape the responses by governments, and vice versa. These factors may influence, for example, the instruments a government may use in addressing a given interop problem. To generalize, European lawmakers have appeared to be more inclined to regulate interop ex ante than their US counterparts, whereas US lawmakers have tended to rely on market forces up front and to turn eventually to corrective ex post mechanisms as needed. Several of the most recent examples that we have studied, including e-health and the smart grid, suggest a possible trend toward convergence between the US and European approaches. Increasingly, blended approaches, where public and private actors work together to establish optimal levels of interop, play an important role on both sides of the Atlantic. And, as demonstrated by our examples from Japan, China, and beyond, such approaches, in addition to innovative strategies, are emerging around the world.

The greatest payoff from the close study of interop ought to be the manner in which it guides our decision making on some of the biggest questions of our increasingly global, interconnected, digital world. It should push us, as individuals and as societies, to acknowledge and address the costs and benefits of deep interconnection among technologies, data, humans, and institutions. We need to understand, too, the implications of the failure of complex systems to work together in optimal fashion. Fundamentally, a deep understanding of interop will help us as we work together, across our many roles and functions in society, to fashion the kind of world in which we wish to live.

ACKNOWLEDGMENTS

Over the past decade, many wonderful students, collaborators, colleagues, and friends from around the world, representing many different disciplines and working at great institutions such as universities, companies, and governments, have informed our work on interop. We have been extremely fortunate to receive guidance and feedback on our proposed theory of interoperability in hundreds of formal and informal conversations, interviews, conferences, and workshops, but also through a growing body of interdisciplinary literature. These exchanges have shaped and sharpened our understanding of interoperability in profound ways, although all errors and omissions are ours alone. We are deeply grateful to all our collaborators.

All-star teams on two sides of the Atlantic conducted, with the two of us, large parts of the research that has informed this book. At the Research Center for Information Law at the University of St. Gallen in Switzerland, we received support from an outstanding team, whose members in the meantime have become scholars in their own right. Daniel Haeusermann, Richard Staeuber, and James Thurman did fantastic work on the first case studies and helped organize a series of inspiring and highly productive interoperability workshops in Appenzell and in Cambridge, Massachusetts. Jan Gerlach helped in many ways in the later stages of book writing. The Swiss National Science Foundation supported important parts of the initial phase of research and collaboration as part of a grant on digital communications and the law.

At the Berkman Center for Internet & Society at Harvard University in Cambridge, our shared intellectual home, we have been blessed with an extraordinary research team led by Caroline Nolan, who is a miracle worker and a truly wonderful colleague. Without Caroline's magic and the extraordinary research support from June Casey at the Harvard Law School Library, we would not have been able to write this book. We owe both of them enormous thanks, in the context of this book and much beyond. We are also thankful for the great help and support we received from other members of our core team. Carey Andersen, Amar Ashar, David O'Brien, Sandra Cortesi, Sebastian Diaz, Rob Faris, Karyn

Glemaud, Dan Jones, Seongmin Lee, Nathaniel Levy, Jon Murley, Rebecca Tabasky, and Seth Young have each contributed substantially to this research project. Our friend Colin Maclay has helped create this amazing environment and, as our collaborator, has helped manage our teams. We dedicate this book to all members of the Berkman Center.

Over the past few years, an extremely talented and inspiring group of research assistants helped produce this book by drafting additional case studies and conducting background research: Jacob Albert, Matthew Becker, Andrew Crocker, Virgina Fuller, Tim Grayson, Adam Holland, Jack Holkeboer, Paul Kominers, James Kwok, Shane Matthews, Ian McClain-Sewer, Amy Rabinowitz, David Russcol, Felix Treguer, and Sally Walkerman. Thanks also to our academic friends at the NEXA Center for Internet & Society at the Politecnico di Torino in Italy and KEIO University in Japan, as well as to our partners from industry, particularly Annmarie Levins and Nick Tsilas at Microsoft, for their very helpful feedback, especially on our initial interop research and writing.

We are especially grateful to our colleagues whose work has inspired this book and who have offered both guidance and helpful critique along the way. Our mentor Terry Fisher provided important feedback on a draft version of this book, and Jonathan Zittrain offered helpful comments at a salon dinner he kindly hosted for us at his home. As with all of our writing, the scholarship of our teachers Yochai Benkler, Lawrence Lessig, and Charles Nesson, as well as Jean Nicolas Druey and Herbert Burkert, has been a constant influence. François Lévêque, Phil Malone, and Viktor Mayer-Schoenberger provided helpful inputs and insightful comments at multiple stages of the research process. Martha Minow, the dean of Harvard Law School, has been a strong supporter of our joint research efforts and the work of the Berkman Center.

After *Born Digital*, we had the great fortune to work, for a second time, with our editor Lara Heimert and her outstanding team at Basic Books. Although it seemed impossible at the outset, she made us work even harder on this book than on the previous one. We could not be more grateful for her editorial guidance and ongoing support. Without a doubt, she is both the best and toughest editor one could wish for. Thanks, too, to Basic Books' wonderful editorial, marketing, and publicity teams.

As ever, our last, and most important, thanks go to our families and closest friends, who have encouraged and patiently supported the creation of another book. Thanks to all of you, individually and collectively. We promise to increase our own human interoperability skills to make up for it in the future.

NOTES

Introduction

1. The case studies as well as additional materials for further exploration are available at http://cyber.law.harvard.edu/interoperability/.

2. Mashups in the web services context are fusions of data from two or more web applications to create an integrated experience informed by the original data sources. Mashup creators pull data dynamically from one source and integrate it with another. As a simple example, Fast Food Maps combines location information of major U.S. fast food restaurants with Google Maps so residents of a particular city can see where they can stop for a hamburger or pizza. See John Palfrey and Urs Gasser, "Mashups Interoperability and eInnovation," case study, Berkman Publication Series, November 2007, http://cyber.law.harvard .edu/interop/pdfs/interop-mashups.pdf.

3. "Cloud computing," which we will discuss in greater detail in the last chapter of the book, is a broad term with several meanings. In simple terms, it describes the practice of using a network of *remote* instead of local servers to store, manage, and process data. Web e-mail services such as Gmail or Hotmail and social networking platforms such as Facebook or Xing are examples of applications "in the cloud," where software is delivered as a service over the Internet rather than shipped as a product that needs to be installed on a local computer or server.

4. Keith Barry, "May 10, 1869: Golden Spike Links Nation by Rail," This Day in Tech: Events That Shaped the Wired World, *Wired*, May 10, 2010, http://www.wired.com/thisdayintech/2010/05/0510transcontinental-railroad -completed/.

Chapter One

1. Servers are computers that link PCs together in networks.

2. An operating system is a complex piece of software that makes computers run their most basic functions.

3. Augmented reality technology allows an overlay of explanatory data on videos or on real-life scenes. See, for example, Charles Arthur, "Augmented Reality: It's Like Real Life, but Better," *Observer*, March 20, 2010, http://www.guardian .co.uk/technology/2010/mar/21/augmented-reality-iphone-advertising.

4. Steven Mostyn, "Third-Generation Kindle Reader Is Amazon's Best-Selling Product," *Tech Herald*, December 28, 2010, http://www.thetechherald.com/article .php/201052/6615/Third-generation-Kindle-reader-is-Amazon-s-best-selling -product.

5. See Jonathan L. Zittrain, *The Future of the Internet—and How to Stop It* (New Haven: Yale University Press, 2008).

6. Many books have taken up this issue of data privacy and the role of aggregators. See, for instance, Daniel Solove, *The Digital Person: Technology and Privacy in the Information Age* (New York: New York University Press, 2004); and Lori Andrews, *I Know Who You Are and I Saw What You Did: Social Networks and the Death of Privacy* (New York: Free Press, 2011).

Chapter Two

1. US Department of Transportation, "2010 Traffic Data for U.S. Airlines and Foreign Airlines U.S. Flights: Total Passengers Up from 2009, Still Below 2008," Bureau of Transportation Statistics data, March 22, 2011, http://bts.gov/press _releases/2011/bts017_11/pdf/bts017_11.pdf.

2. US Centennial of Flight Commission, "Air Traffic Control," http://centen nialofflight.gov/essay/Government_Role/Air_traffic_control/POL15.htm.

3. International Civil Aviation Organization, "Flight Safety (FLS) Section, Personnel Licensing—Frequently Asked Questions," http://legacy.icao.int/icao /en/trivia/peltrgFAQ.htm.

4. See generally European Commission, "Interoperability Solutions for European Public Administrations," last revised January 1, 2012, http://ec.europa .eu/isa/; European Commission, "Communication from the Commission to the European Parliament, the Council, the European Economic Social Committee and the Committee of the Regions: Towards Interoperability for European Public Services," Brussels, COM (2010) 744 final, December 16, 2010, http:// ec.europa.eu/isa/documents/isa_iop_communication_en.pdf.

Chapter Three

1. International Telecommunication Union (ITU), "Universal Phone Charger Standard Approved," press release, October 22, 2009, http://www.itu.int /newsroom/press_releases/2009/49.html.

2. NPD Group, "The NPD Group: Amazon Ties Walmart as Second-Ranked U.S. Music Retailer, Behind Industry-Leader iTunes," press release, May 26, 2010, http://www.npd.com/press/releases/press_100526.html.

3. International Federation of Phonographic Industry, "IFPI Publishes Digital Music Report 2011," January 20, 2011, http://www.ifpi.org/content/section _resources/dmr2011.html.

4. As we will later discuss in Chapter 5, such practices are likely to reduce competition, at least in the long run, and have also triggered strong reactions by regulators in Europe.

5. Peter Cohen, "iTunes Store Goes DRM-Free," *MacWorld.com*, January 6, 2009, http://www.macworld.com/article/137946/2009/01/itunestore.html/. See also iTunes Store: iTunes Plus Frequently Asked Questions (FAQ), last modified June 25, 2010, http://support.apple.com/kb/ht1711.

6. Digital Entertainment Content Ecosystem, "What Is UltraViolet?" 2011, http://www.uvvu.com/what-is-uv.php.

7. M. G. Siegler, "With DECE's UltraViolet, We're About to See Just How Powerful Apple Really Is," *TechCrunch*, July 20, 2010, http://techcrunch.com /2010/07/20/dece-ultraviolet-apple.

8. See UltraViolet Service Privacy Policy, September 2, 2011, https://my .uvvu.com/ssp/public/privacyStatementPage.jsf.

9. A "client" is a software application that communicates and exchanges data with a server that is, typically, accessible over a network, such as a local area network (LAN) or the Internet. In this example, a user accesses the Tencent QQ instant messaging services over the Internet by using the QQ software locally installed on the user's computer.

10. Pingdom, "Amazing Facts and Figures About Instant Messaging (Infographic)," *Royal Pingdom* (blog), April 23, 2010, http://royal.pingdom.com /2010/04/23/amazing-facts-and-figures-about-instant-messaging-infographic.

11. As in the case of many highly technical matters, the Wikipedia entry provides a reliable and detailed account. See http://en.wikipedia.org/wiki/Extensible _Messaging_and_Presence_Protocol (accessed January 1, 2012).

12. Amanda Lenhart, Kristen Purcell, Aaron Smith, and Kathryn Zickur, "Social Media and Mobile Internet Use Among Teens and Young Adults," *Pew Internet and American Life Project*, February 3, 2010, a project for the Pew Research Center, http://www.pewinternet.org/~/media//Files/Reports/2010/PIP _Social_Media_and_Young_Adults_Report_Final_with_toplines.pdf.

13. Alicia Ashby, "$3.9 Bill in Global Revenue from 500 Virtual Worlds in 2011, Reports KZero," *Engage Digital*, February 2, 2011, http://www.engage digital.com/blog/2011/02/02/kzero-39b-global-revenue-from-virtual-worlds -in-2011/.

Chapter Four

1. Google has also used interoperability as a way of giving its customers more visibility into the data collected about them. A service called Dashboard enables Google users to see most of the forms of information Google has collected about them. The Dashboard service is intended to provide greater transparency to users and to help them make better choices about what to disclose to Google.

2. Ylan Q. Mui and Pete Whoriskey, "Facebook Cements No. 1 Status," *Washington Post*, December 31, 2010, http://www.washingtonpost.com/wp-dyn /content/article/2010/12/30/AR2010123004625.html.

3. See the official Google blog posts about Buzz, including its introduction, which references interoperability obliquely as well as Google's response to the

negative user experiences: Todd Jackson, "Introducing Google Buzz," *Official Google Blog*, February 9, 2010, http://googleblog.blogspot.com/2010/02/introducing -google-buzz.html; Todd Jackson, "A New Buzz Start-Up Experience Based on Your Feedback," *Official Google Blog*, February 13, 2010, http://gmailblog.blog spot.com/2010/02/new-buzz-start-up-experience-based-on.html.

4. By way of disclosures: one of the authors, Urs Gasser, serves on the advisory board for EPIC. Also, both of the authors are colleagues of Professor Rubenstein at Harvard Law School. The settlement of the litigation resulted in a cy pres award of funding to the Berkman Center for Internet and Society at Harvard University, with which the authors are associated. Neither author was directly involved in this litigation.

5. See Consolidated and Amended Class Action Complaint, *In re Google Buzz User Privacy Litigation*, Case no. 5:10-cv-00672-JW (N.D. Cal. July 30, 2010). (The official website is no longer available since the settlement was approved, but it was located at http://www.buzzclassaction.com/. The former website can be accessed via Internet Archive, http://www.archive.org/.)

6. David Kravets, "Judge Approves $9.5 Million Facebook 'Beacon' Accord," *Wired*, March 17, 2010, http://www.wired.com/threatlevel/2010/03/facebook -beacon-2.

7. See Federal Trade Commission, "In the Matter of Google, Inc., a Corporation," FTC file no. 102 3136, October 24, 2011, http://www.ftc.gov/os/caselist /1023136/index.shtm.

Chapter Five

1. Thomas Crampton, "France Approves 'iPod Law' on Music Downloads," *International Herald Tribune*, June 30, 2006, http://www.nytimes.com/2006/06 /30/technology/30iht-copy.2093092.html. See also "France Gives Apple a Break in Interoperability Case," *CIO.com*, July 31, 2006, http://www.cio.com/article /23470/France_Gives_Apple_a_Break_in_iTunes_Interoperability_Case.

2. Viewed from this angle, the behavior of many high-tech companies that collapsed during the first dot-com bubble in 2000, after having invested in "communities" and brand awareness by giving away services and products for free, seems understandable.

3. For an overview, see Nils Brunsson, "Standardization and Uniformity," in *A World of Standards*, by Nils Brunsson and Bengt Jacobsson (New York: Oxford University Press, 2002), 138–150.

4. See, for example, Murray Gell-Mann, *The Quark and the Jaguar: Adventures in the Simple and the Complex* (London: Abacus, 1994), 329–344.

Chapter Six

1. See Zittrain, *The Future of the Internet*, for an extensive discussion of the concept of generativity. Zittrain defines generativity as "[a] technology's overall

capacity to produce unprompted change driven by large, varied, and uncoordinated audiences." Zittrain, "The Generative Internet," *Harvard Law Review* 119 (May 2006): 1974, 1980.

2. *Ushahidi* means "testimony" in Swahili.

3. Geocoding is the process of converting an informal location identifier, such as a street address, into formal geographic coordinates.

4. An application programming interface (API) serves as an open connection among different software programs and makes possible their interaction.

5. See, for example, Programmable Web's directory of the most popular APIs: http://www.programmableweb.com/apis/directory/1?sort=mashups.

6. See, for example, Jon Jenson, "A Region in Upheaval: First Tunisia, Now Bahrain? As Unrest Spreads, Here's What You Need to Know," *GlobalPost*, January 27, 2011, http://www.globalpost.com/dispatch/africa/110126/protests -riots-tunisia-egypt-lebanon-middle-east-north-africa.

7. "Twitter Registers 1,500 Per Cent Growth in Users," *New Statesman*, March 4, 2010, http://www.newstatesman.com/digital/2010/03/twitter-registered -created.

8. For the official press release from Facebook, see "Facebook Unveils Platform for Developers of Social Applications," May 24, 2007, https://www .facebook.com/press/releases.php?p=3102. For a recent analysis, see Gwange Jung and Byngtae Lee, "How Did Facebook Outpace Myspace with Open Innovation? An Analysis of Network Competition with Changes of Network Topology," *PACIS 2011 Proceedings*, paper 88, July 9, 2011, http://aisel.aisnet.org/cgi /viewcontent.cgi?article=1087&context=pacis2011.

9. A wiki is a website that allows users to easily and collaboratively edit or otherwise modify content displayed on its web pages through a web browser. Although many wikis exist on the Internet, perhaps the best-known example is Wikipedia. For more information on wikis and wiki software, see the Wikipedia entry "Wiki," http://en.wikipedia.org/wiki/Wiki. The term "crowdsourcing" is a portmanteau of "crowd" and "outsourcing" that is often used to describe a distributed problem-solving and production model whereby tasks that are traditionally performed by specifically trained individuals are instead collaboratively performed by an open community comprised of "everyday people." See Jeff Howe, "The Rise of Crowdsourcing," *Wired*, June 2006, http://www.wired.com /wired/archive/14.06/crowds.html.

10. John Palfrey and Urs Gasser, "Case Study: Mashups Interoperability and eInnovation," November 2007, Berkman Center Research Publication no. 2007-10, SSRN, http://www.ssrn.com/abstract=1033232.

11. Jeffrey K. MacKie-Mason and Janet S. Netz, "Manipulating Interface Standards as an Anticompetitive Strategy," in *Standards and Public Policy*, ed. Shane Greenstein and Victor Stango (New York: Cambridge University Press, 2007), 231–259, on 255.

12. As used here, "information communications technologies (ICT)" broadly describes the technologies, systems, and networks that enable communication and access to information, knowledge, and entertainment through the Internet, telecommunications, cellular wireless networks, and other media.

13. Gaia Bernstein, "In the Shadow of Innovation," *Cardozo Law Review* 31 (2010): 6.

Chapter Seven

1. See Jeremy Landt, "The History of RFIDs," *IEEE Potentials,* October–November 2005, 8.

2. See Mark Roberti, "The History of RFID Technology," *RFID Journal,* n.d., http://www.rfidjournal.com/article/view/1338.

Chapter Eight

1. For an overview, see Donella Meadows, "Leverage Points: Places to Intervene in a System" (Hartland, VT: Sustainability Institute, 1999), http://www .sustainer.org/pubs/Leverage_Points.pdf.

2. For an in-depth review of the connections between open standards and interop, see Laura DeNardis, *Opening Standards: The Global Politics of Interoperability* (Cambridge: MIT Press, 2011).

Chapter Nine

1. MGM Studios, Inc. v. Grokster, Ltd., 545 U.S. 913 (2005).

2. Brad Biddle, Andrew White, and Sean Woods, "How Many Standards in a Laptop? (and Other Empirical Questions)," September 10, 2010, SSRN, http:// ssrn.com/abstract=1619440.

3. For a thorough and illuminating study of open standards processes and related questions of technology governance, see DeNardis, *Opening Standards.*

4. ODF is a file format for representing electronic documents such as word processing texts, spreadsheets, and presentations.

Chapter Ten

1. An especially thoughtful take on the relationship between law and interoperability can be found in Jonathan Band and Masanobu Katoh, *Interfaces on Trial 2.0* (Cambridge: MIT Press 2011).

2. See "Which Is the Legal Framework for Technological Protection Measures?" World Intellectual Property Organization, http://www.wipo.int/enforce ment/en/faq/technological/faq04.html.

3. See John Markoff, "Vast Spy System Loots Computers in 103 Countries," *New York Times,* March 28, 2009, http://www.nytimes.com/2009/03/29 /technology/29spy.html?pagewanted=all. Ronald Deibert, Rafal Rohozinski, Nart Villeneuve, and Greg Walton were the researchers who uncovered this spy

ring and published a groundbreaking report under the banner of the Information Warfare Monitor.

Chapter Eleven

1. See James H. Bigelow, Katya Fonkych, and Federico Girosi, "Technical Executive Summary in Support of 'Can Electronic Medical Record Systems Transform Healthcare?' and 'Promoting Health Information Technology'" (RAND Health working paper WR-295, September 14, 2005), http://www.rand.org /content/dam/rand/pubs/working_papers/2005/RAND_WR295.pdf.

2. K. C. Jones, "Obama Wants E-Health Records in Five Years," *Information Week*, January 12, 2009, http://www.informationweek.com/news/healthcare /212800199. See also K. C. Jones, "U.S. House Pushes for National E-Health Records," June 26, 2008, http://www.informationweek.com/news/software /info_management/208801143.

3. Nicholas Timmons, "Only the Bare Bones," *Financial Times*, May 17, 2011, 9.

4. See OECD, "Health at a Glance 2011: OECD Indicators. Key Findings: United States," 1, http://www.oecd.org/dataoecd/12/58/49084319.pdf.

5. See Robert Wood Johnson Foundation, "About the Commission to Build a Healthier America," http://www.rwjf.org/files/research/commissionone pager.pdf. For the full report and further research, see Robert Wood Johnson Foundation Commission to Build a Healthier America, "Beyond Health Care: New Directions to a Healthier America," http://www.commissiononhealth.org.

6. The RAND Corporation is "a nonprofit institution that helps improve policy and decisionmaking through research and analysis." RAND Corporation, http://www.rand.org. For a summary of RAND research findings, see RAND Corporation, "Health Information Technology—Can HIT Lower Costs and Improve Quality? Research Highlights" (RAND research brief RB-9136, 2005), http://www.rand.org/pubs/research_briefs/RB9136/index1.html. For a more comprehensive article, see Richard Taylor, Anthony Bower, Federico Girosi, James Bigelow, Katya Fonkych, and Richard Hillestad, "Promoting Health Information Technology: Is There a Case for More-Aggressive Government Action?" *Health Affairs* 24, no. 5 (September 14, 2005): 1234–1245, http:// content.healthaffairs.org/content/24/5/1234.full.

7. Richard Hillestad, James Bigelow, Anthony Bower, Federico Girosi, Robin Meili, Richard Scoville, and Roger Taylor, "Can Electronic Medical Record Systems Transform Health Care? Potential Health Benefits, Savings, and Costs," *Health Affairs* 24, no. 5 (September 14, 2005), 1108, http://www.eecs.harvard .edu/cs199r/readings/RAND_benefits.pdf.

8. Federico Girosi, Robin Meili, and Richard Scoville, *Extrapolating Evidence of Health Information Technology Savings and Costs* (Santa Monica, CA: RAND Corporation, 2005); text available online at http://www.rand.org/pubs

/monographs/MG410.html; Hillestad et al., "Can Electronic Medical Record Systems Transform Health Care?"

9. Congressional Budget Office, "Evidence on the Costs and Benefits of Health Information Technology," May 2008, http://www.cbo.gov/ftpdocs/91xx /doc9168/05-20-healthit.pdf.

10. Andrew King, "How Kaiser Permanente Went Paperless," *Bloomberg Businessweek,* April 7, 2009, http://www.businessweek.com/technology/content /apr2009/tc2009047_562738.htm.

11. For an overview of the theory of generativity in the context of information technology systems, see Zittrain, *The Future of the Internet.*

12. US Department of Health and Human Services, "Health Information Privacy—The Privacy Rule," http://www.hhs.gov/ocr/privacy/hipaa/administrative /privacyrule/index.html.

13. See US Department of Health and Human Services, "Modifications to the HIPAA Privacy, Security, and Enforcement Rules Under the Health Information Technology for Economic and Clinical Health Act," *Federal Register* 75, no. 134 (July 14, 2010): 40,868, http://www.hhs.gov/ocr/privacy/hipaa /understanding/coveredentities/nprmhitech.pdf.

14. Drs. Kenneth D. Mandl and Isaac S. Kohane make the case for the concept of substitutability in Mandl and Kohane, "No Small Change for the Health Information Economy," *New England Journal of Medicine* 360, no. 13 (March 26, 2009): 1278–1281; online at http://www.nejm.org/doi/full/10.1056/NEJMp 0900411.

15. Janet Adamy, "U.S. Eases Funds to Adopt Electronic Medical Records," *Wall Street Journal,* July 14, 2010, http://online.wsj.com/article/SB1000 1424052748703834604575365413124717600.html. For a description of the new rules, see this essay on the *New England Journal of Medicine*'s website: David Blumenthal and Marilyn Tavenner, "The 'Meaningful Use' Regulation for Electronic Health Records," *New England Journal of Medicine,* July 13, 2010, http:// healthcarereform.nejm.org/?p=3732&query=home.

16. Julia Adler-Milstein, David W. Bates, and Ashish K. Jha, "U.S. Regional Health Information Organizations: Progress and Challenges," *Health Affairs* 28, no. 2 (March–April 2009): 483–492, http://content.healthaffairs.org/content /28/2/483.abstract.

17. Such a system might emerge as a matter of practice among private-sector actors, but that seems unlikely. An example of private ordering of this sort is the payment among libraries for cataloging records through a collaborative, nonprofit service called OCLC. Libraries that create more and better records that are copied by other institutions can receive rebates on the costs of using the system. A small incentive is established to encourage each library to do a more extensive job of cataloging books. In this case, a stronger nudge from the government, and a bigger incentive, appears to be necessary.

18. See Ashish K. Jha and David Blumenthal, "International Adoption of Electronic Health Records," in Robert Wood Johnson Foundation, George Wash-

ington University Medical Center, and Institute for Health Policy, "Health Information Technology in the United States: Where We Stand, 2008," 104–128, http://www.rwjf.org/files/research/3297.31831.hitreport.pdf.

19. Eben Harrell, "In Denmark's Electronic Health Records Program, a Lesson for the U.S.," April 16, 2009, *TIME*, http://www.time.com/time/health/article/0,8599,1891209,00.html.

20. Lene Grosen, "Electronic Health Record in Denmark." *Health Policy Monitor*, October 2009, http://www.hpm.org/survey/dk/a14/5.

Chapter Twelve

1. See, for example, John Bailey's examination of the effects of the slow fire phenomenon in the context of holdings at the Morgan Library: Bailey, "'Slow Fires' and the Morgan Library," *John's Bailiwick* (blog), October 23, 2009, American Society of Cinematographers, http://www.ascmag.com/blog/2009/10/23/%E2%80%9Cslow-fires%E2%80%9D-and-the-morgan-library.

2. Consider the nonfiction books by Jonathan Zittrain (*The Future of the Internet*), Yochai Benkler (*The Wealth of Networks*, 2006), and Lawrence Lessig (*Code and Other Laws of Cyberspace 2.0*, 2006; *The Future of Ideas*, 2001; *Free Culture*, 2004; and *Remix*, 2008), which have been published freely online, in html-based versions, and have sold well as printed works.

3. See, for example, Brian Lavoie, "Meeting the Challenges of Digital Preservation: The OAIS Reference Model," *OCLC Newsletter*, no. 243 (January–February, 2000): 26–30, as it appears at http://www.oclc.org/research/publications/archive/2000/lavoie/.

4. Jacqui Cheng, "Publishers Want Universal E-Books; Won't Cooperate to Get Them," *Ars Technica*, n.d. [c. June 2010?], http://arstechnica.com/gadgets/news/2010/06/publishers-want-universal-e-books-cant-cooperate-to-get-them.ars.

5. See Legal Information Preservation Alliance, http://www.aallnet.org/lipa/; Law Library Microform Consortium, "About LIPA," http://www.llmc.com/AboutLIPA.asp.

6. See Nicholson Baker, *Double Fold: Libraries and the Assault on Paper* (Random House, 2001). For a response, see Richard J. Cox, "Don't Fold Up: Responding to Nicholson Baker's *Double Fold*," Society of American Archivists, April 18, 2001, http://www.archivists.org/news/doublefold.asp.

Chapter Thirteen

1. On matters related to technology, Wikipedia entries are often excellent sources. The cloud computing entry, as of January 12, 2012, is excellent. See http://en.wikipedia.org/wiki/Cloud_computing.

2. National Institute of Standards and Technology, Information Technology Laboratory, "Cloud Computing Forum and Workshop," updated August 20, 2011, http://csrc.nist.gov/groups/SNS/cloud-computing.

3. Microsoft, "Case Study: City of Miami—City Government Improves Service Offerings, Cuts Costs, with 'Cloud' Services Solution," February 24, 2010, http://

www.microsoft.com/casestudies/case_study_detail.aspx?casestudyid=400000
6568.

4. Narendran Calluru Rajasekar, "Web Services—Current Trends and Future
Opportunities," November 19, 2009, University of East London, http://www
.narensportal.com/papers/web-services-future-opportunities.aspx.

5. Vivek Kundra, "Federal Cloud Computing Strategy," February 8, 2011, 7,
http://www.cio.gov/documents/Federal-Cloud-Computing-Strategy.pdf.

6. Fred Humphries, "Congress Considers Cloud Computing," *TechNet Blogs:
Microsoft on the Issues*, September 22, 2010, http://blogs.technet.com/b
/microsoft_on_the_issues/archive/2010/09/22/congress-considers-cloud
-computing.aspx.

7. A. J. Watson et al., "Diabetes Connected Health: A Pilot Study of Patient-
and Provider-Shared Glucose Monitoring Web Application," *Journal of Diabetes
Science and Technology* 3, no. 2 (March 2009): 345–352, http://www.connected
-health.org/programs/diabetes/research-materials-external-resources/diabetes
-connected-health-a-pilot-study-of-a-patient-and-provider-shared-glucose-mon
itoring-web-application.aspx.

8. Catherine Tsai, "Lab Hopes Building Makes More Power Than It Uses,"
Associated Press, February 23, 2011; Kirk Johnson, "Soaking Up the Sun to
Squeeze Bills to Zero," *New York Times*, February 14, 2011.

9. Office of the National Coordinator for Smart Grid Interoperability, "NIST
Framework and Roadmap for Smart Grid Interoperability Standards, Release
1.0," NIST Special Publication 1108, January 2010, http://www.nist.gov/public
_affairs/releases/upload/smartgrid_interoperability_final.pdf.

10. National Institute of Standards and Technology, "Smart Grid Panel Agrees
on Data-Exchange Standards for Electricity Usage," February 17, 2011, http://
www.nist.gov/smartgrid/smartgrid-021711.cfm.

11. University of Southern California, School of Cinematic Arts, "Million
Story Building," http://cinema.usc.edu/interactive/research/millionstory.cfm.

12. See Jennifer Stein, "Mobile and Ambient Storytelling," Mobile and Environmental Media Lab—University of Southern California, http://jenstein.net
/MSB/MillionStory_paper_USC.pdf.

13. See, for example, Claire Swedberg, "Wristbands Document Interactions
Between Prisoners and Officers," *RFID Journal*, March 4, 2010, http://www
.rfidjournal.com/article/view/7436.

14. Chris Cameron, "Forget Hall Monitors, School Investigates Tracking Students with RFID," *Read Write Web*, August 23, 2010, http://www.readwrite
web.com/archives/forget_hall_monitors_school_investigates_tracking_students
_rfid.php.

15. Elgar Fleisch, "What Is the Internet of Things? An Economic Perspective,"
Economics, Management and Financial Markets 5, no. 2 (2010): 125–157, 133.

SUGGESTED READINGS

Since we began conducting research on interconnected systems almost a decade ago, we have learned from a rich and diverse set of materials from various disciplines—ranging from the highly theoretical to the narrowly pragmatic—as we have developed our theory of interoperability in the digital age. Our study has focused on those materials that help develop a deeper understanding of how and when interop works and why it may fail in its application. We cite a relatively small number of works in the text, in order to keep the reading experience streamlined, but even those few notes point to the wide range of work that has already gone into seeking to understand this phenomenon. The materials cited in the notes have deeply shaped our thinking about the multifaceted interoperability problem (and opportunity, for that matter). They are also an excellent starting point for readers who want to dig deeper into specific subject matter areas we have touched upon in this book.

In addition to the materials listed in the endnotes throughout this book, the interested reader will find links, pointers, and references to a great variety of materials on the interoperability wiki we have developed with our research team at the Berkman Center for Internet and Society at Harvard University. The wiki can be accessed at http://cyber.law.harvard.edu/interoperability. From there, we point to additional background readings as well as to our own research, the basis on which we have written this book, including our 2007 white paper and corresponding case studies on ICT interoperability and innovation. Further, the wiki also features a series of fascinating use cases and mini–case studies drafted by a group of talented students on our research team. These case studies provide in-depth insights into interoperability issues in such diverse areas as currency, railway systems, the smart grid, the Internet of Things, and bar codes, to mention just a few. Each of these documents includes many more specific references to further readings and for context-specific exploration of various interop problems. Together, these materials build a virtual "second layer" of this book and offer a deeper way of engaging with the core theories and issues we consider here.

Our research has built on the shoulders of giants from various disciplines, including economics, law, communications, computer science, engineering, political science, psychology, sociology, and so forth. Although the work of many wonderful authors and researchers has influenced our thinking, we would like to highlight a few readings that we highly recommend to everyone who would like to think further about the promise and challenges of interconnected systems in the digital world.

The first set of recommended readings deals with various theoretical and practical aspects as well as with the normative implications of the *innovative power of the Internet.* Yochai Benkler's *Wealth of Networks* (2006), Jonathan Zittrain's *The Future of the Internet and How to Stop It* (2008), Barbara van Schewick's *Internet Architecture and Innovation* (2010), and Eric von Hippel's *Democratizing Innovation* (2005) on user innovation have framed much of our thinking on the crucial role that interoperability plays as an enabler of innovation in the digitally networked world. Our belief in the deeply human relevance of interoperability—and the important role law can and should play in facilitating it—has been shaped by William Fisher's work on semiotic democracy, *Promises to Keep: Technology, Law, and the Future of Entertainment* (2004), and his vision of human flourishing, "Implications for Law of User Innovation" (*Minnesota Law Review* 94, no. 5 [May 2010]: 1417–1477), as well as by Lawrence Lessig's seminal work on technology and culture, *Free Culture: How Big Media Uses Technology and the Law to Lock Down Culture and Control Creativity* (2004).

In addition to these foundational pieces, we have greatly benefited from the many wonderful books and articles written by colleagues from other schools and departments addressing *specific elements* that are at the core of our theory of interoperability. Although it is impossible to do justice to the many great sources of inspiration and insight, we would like to highlight just a few pieces that might serve as helpful starting points for the reader to start his or her own deep dives.

As an introduction to ICT and interoperability, the policy essays in the collection edited by Laura DeNardis, *Opening Standards: The Global Politics of Interoperability* (2011), provide a general overview, including a good introduction by John B. Morris Jr. in his "Injecting the Public Interest into Internet Standards" and by Andrew Updegrove in his "ICT Standards Setting Today: A System Under Stress."

Approaching interoperability through the lens of *economics*, we suggest Hal Varian, Joseph Farrell, and Carl Shapiro's *The Economics of Information Technology* (2004) as a good general stage-setter in the digital technology environment. On the specific topics of competition, network effects, standardization, lock-ins, and related issues, we can recommend as background readings—with different focus areas, examples, and orientation—Michael L. Katz and Carl Shapiro,

"Systems Competition and Network Effects" (*Journal of Economic Perspective* 8, no. 2 [Spring 1994]: 93–115); Neil Gandal, "Compatibility, Standardization, and Network Effects: Some Policy Implications" (*Oxford Review of Economic Policy* 18, no. 1 [2002]: 80–91); Paul A. David and Shane Greenstein, "The Economics of Compatibility Standards: An Introduction to Recent Research" (*Economics of Innovation and New Technology* 1 [1990]: 3–41); S. J. Liebowitz and Stephen E. Margolis, "Path Dependence, Lock-In, and History" (*Journal of Law, Economics, and Organization* 11, no. 1 [1995]: 205–226); and Chapter 17, "Network Effects," in David Easley and Jon Kleinberg's *Networks, Crowds, and Markets: Reasoning About a Highly Connected World* (2010).

A diverse and rapidly growing body of scholarship from the disciplines of *law and public policy* is relevant to the emerging theory and practice of interoperability as outlined in this book. Among the many things we have read on the law and policy of interoperability, Stacy A. Baird's "Government Role and the Interoperability Ecosystem" (*I/S: A Journal of Law and Policy* 5, no. 2 [2009]: 219–290) gets close to some of the general arguments we have tried to build over the years. Viktor Mayer-Schoenberger's work on legal and policy issues around interoperability, such as "Emergency Communications: The Quest for Interoperability in the United States and Europe" (Kennedy School of Government Faculty Research Working Papers Series RWP02-024, March 2002), is also a must-read on this topic, as are the contributions and case studies in Shane Greenstein and Victor Stango, eds., *Standards and Public Policy* (2007).

In addition to interop-oriented scholarship, knowledge in *specific areas of law*—including, for instance, intellectual property (IP) law and competition law—is key to further discuss the arguments presented in this book. The paradigm shift in the way we (should) think about IP rights and how we exercise them in the digital age has been recently discussed in John Palfrey's *Intellectual Property Strategy* (2011) and in Ben Klemens's *Math You Can't Use: Patents, Copyright, and Software* (2006), books that will also help the reader understand our argument on the difficult relation between IP and interoperability. More specific readings for further research into the interplay between IP and competition law and its effects on interoperability include Jonathan Band and Masanobu Katoh's *Interfaces on Trial 2.0* (2011), as well as various accounts of the Microsoft cases, including William H. Page and John E. Lopatka, *The Microsoft Case: Antitrust, High Technology, and Consumer Welfare* (2007), or the articles by Christian Ahlborn and David S. Evans, "The Microsoft Judgment and Its Implications for Competition Policy Towards Dominant Firms in Europe" (*Antitrust Law Journal* 75, no. 3 [2009]: 887–932), or Claudia Schmidt and Wolfgang Kerber, "Microsoft, Refusal to License Intellectual Property Rights, and the Incentives Balance Test of the EU Commission" (November 8, 2008, available at SSRN: http://ssrn.com/abstract=1297939). As far as the question of legal interoperability is concerned, we recommend Dan L. Burk, "Law as a Network Standard" (*Yale Journal of Law and Technology* 8 [2005]: 63–77), and Margaret Jane Radin, "Online Standardization and the Integration of Text and Machine" (*Fordham Law Review* 70, no. 4 [2002]: 1125–1146), as interesting readings with further references. Cass R. Sunstein discusses in his article "Incompletely Theorized Agreements in Constitutional Law" (*Social Research* 74, No 1 [2007]: 1–24) a sophisticated legal mechanism that has emerged in the field of constitutional law and helps us to achieve workable levels of interoperability despite disagreements about the theories based on which we make our value judgments.

Much of the scholarship we recommend for further reading and exploration (including some of the pieces already mentioned) transcends disciplinary boundaries. In this context, consider the important strand of recent research on *human cooperation*, a subject that we have only tangentially covered in this book but that is of crucial importance to gaining a better future understanding of interoperability at the human and institutional layers. A great introduction and vision was recently published by our colleague Yochai Benkler, *The Penguin and the Leviathan: The Triumph of Cooperation over Self-Interest* (2011). Similarly, an enormously rich body of research on *complexity and complex systems* shapes our understanding of the promise and perils of highly interconnected systems. Good and accessible introductions include Murray Gell-Mann's *The Quark and the Jaguar* (1994) and, as a much shorter read, but also very insightful, Steven H. Strogatz's "Exploring Complex Networks" (*Nature* 410, no. 6825 [March 8, 2001]: 268–276) as well as (with an eye toward the question of manageability of complex systems) Donella Meadows's "Leverage Points: Places to Intervene in a System" (Sustainability Institute, 1999). Luis Camarinha-Matos and Hamideh Afsarmanesh examine complex enterprise systems in "Collaborative Networks: A New Scientific Discipline" (*Journal of Intelligent Manufacturing* 16, nos. 4–5 [2005]: 439–452).

Last but not least, we would like to draw reader's attention to a series of *field-specific studies* of interoperability in practice that we found fascinating. Many more references are listed in the student-written case studies we mentioned above, but the following selection of writings might be helpful. A wonderful account of the role of interoperability in transportation systems is provided by Marc Levinson's *The Box: How the Shipping Container Made the World Smaller and the World Economy Bigger* (2006), which looks at how standard-sized shipping containers have revolutionized global trade. Another intriguing case study on transportation interoperability is provided in Douglas Puffert's *Tracks Across Continents, Paths through History: The Economic Dynamics of Standardization in Railway Gauge* (2009). E-Government is another interesting and illustrative case study where interoperability plays a key role, as the materials in the context of the European Interoperability Framework (EIF) initiative illustrate; see, for example, Luis Guijarro-Coloma's "ICT Standards and Public Procurement in the United States and in the European Union: Influence on eGovernment Deployment," (*Telecommunications Policy* 33, nos. 5–6 [June 2009]: 285–295). An emerging area of interop discussion is found in the medical literature related to electronic health records exchange, works such as Kenneth Mandl and Isaac Kohane, "No Small Change for the Health Information Economy" (*New England Journal of Medicine* 26, no. 13 [March 26, 2009]: 1278–1281), or Rita Kukafka et al., "Redesigning Electronic Health Record Systems to Support Public Health" (*Journal of Biomedical Informatics* 40, no. 4 [August 2007]: 398–409). Another relatively new area of interop study is found in the literature of smart sensor networks, or the Internet of Things, such as the interesting study by Koen Kok and others, "Field-Testing Smart Houses for a Smart Grid" (*Proceedings of the 21st International Conference on Electricity Distribution CIRED*, Frankfurt, 2011), or in the context of network security, as in Sarfraz Alam and Mohammad Chowdhury's "Interoperability of Security-Enabled Internet of Things" (*Wireless Personal Communication* 61 [2011]: 567–586). With respect to cloud computing, as well as to personal computing in the digital era generally, one might consider Frank Gillett's "The Personal Cloud" (July 6, 2009, Forrester Research).

These reading suggestions are meant as a starting point for further research. There is terrific richness in the materials that are directly relevant to the emerging field of interoperability. A devoted reader will find that it takes some work to cross the disciplines that have covered the broad topic of interop from many different angles. But, as we found, it is work well worth doing as we all seek to understand the complexities of the global economy in a digital age.

INDEX